QUANTUM
MECHANICS

QUANTUM
MECHANICS
An Introduction

Dennis Morris

MERCURY LEARNING AND INFORMATION
Dulles, Virginia
Boston, Massachusetts
New Delhi

Publisher: David Pallai
MERCURY LEARNING AND INFORMATION
22841 Quicksilver Drive
Dulles, VA 20166
info@merclearning.com
www.merclearning.com
1-800-232-0223

This book is printed on acid-free paper.

Dennis Morris. *Quantum Mechanics: An Introduction.*
ISBN: 978-1-9422707-9-9

Library of Congress Control Number: 2016935949

161718321 Printed in the United States of America

CONTENTS

INTRODUCTION

Quantum mechanics is not the easiest part of a physics curriculum. Nor is quantum mechanics the easiest subject upon which to write an introductory text book. This is partly because quantum mechanics uses mathematical concepts with which students are often not familiar prior to their introduction to quantum mechanics and partly because it is difficult to present quantum mechanics in a step by step fashion. Most topics of physics can be presented by starting at A, following by B, progressing to C, and ending at Z, all in the proper order. Quantum mechanics is different, and I have seen no text that presents quantum mechanics in such a step by step way. There is also a great proliferation of terminology; most things in quantum mechanics have several names, and it is confusing and difficult to become familiar with all the different names while simultaneously trying to understand the subject. Then there are unintuitive concepts like wavefunction collapse or superposition with which the student has to wrestle.

This book is the 6th draft of my attempt to write an introduction to quantum mechanics that presents the subject simply. Initially, I followed the standard pattern that students will find in the many other quantum mechanics texts. In these texts, the authors go straight to the Schrödinger equation without covering the necessary mathematics to understand it; the necessary mathematics is usually appended to the Schrödinger equation in later chapters (or, often, in other books). Finally, we meet angular momentum and spin. Other material is scattered throughout the texts almost at random. I do not contemn the authors of such texts; I am fully familiar with the difficulty of presenting quantum mechanics in any other way, and for some students, these other authors might well have done a better job than have I in presenting the subject. Now, in my 6th draft, I have concluded that quantum mechanics cannot be presented entirely in a simple step by step way. However, it can be presented more simply than it often is presented.

A look through the contents pages of this book will show that we do not approach the Schrödinger equation until we are more than half way through the book. Indeed, we present an overview of quantum mechanics before we present the Schrödinger equation.

The reader will note that we explain much of the mathematics that underpins the Schrödinger equation within the first part of the book. We deal with angular momentum and intrinsic spin towards the end of the book, and, in this, we follow the normal order of presentation. Although unconventional in the presentation, I hope I have been able in this book to make quantum mechanics less complicated and less difficult to learn than it has the reputation of being.

There is a nagging feeling about quantum mechanics that was shared by many of the greatest physicists of the 20th century and is still felt today by many physicists. That nagging feeling is that there is something about quantum mechanics that is wrong or incomplete. As a physical theory, quantum mechanics has passed every test set for it with 100% success. Yet still, that nagging feeling will not leave the hearts of physicists. We feel that we do not really understand it. Unlike most introductory texts, we do not eschew this nagging feeling. We do not discuss the philosophical aspects of quantum mechanics at great length; that is not the place of an introductory text; but we do give some space over to the philosophical implications of quantum mechanics. In particular, we wonder if the whole mathematical structure of quantum mechanics can somehow be changed to rewrite it in more conventional mathematics (division algebras), and we wonder if the different geometric spaces associated with the different division algebras will allow a sensible extirpation of that nagging feeling.

There is some material in this book that your author has not seen presented elsewhere. In particular, the derivation of the momentum operator as no more than differentiation with respect to the imaginary variable and the derivation of Planck's constant as no more than the inverse of the scaling parameter of the complex numbers, $\mathbb{C}$. Also previously unseen by your author in an introductory textbook is the relativistic derivation of the Schrödinger equation.

There is more to quantum mechanics than can be introduced in a textbook of this size; quantum field theory comes immediately to mind. Preparation for this more advanced material is often omitted from introductory texts. To your author's surprise, restructuring the usual presentation of quantum mechanics allows at least some preparation for the more advanced material, and so there is some such preparation included in this book, but we avoid detailed scrutiny of this advanced material – this book is an introduction only.

Scattered throughout the book are numerous asides. The nature of the asides is varied; some are biographical or historical giving a flavour of the kind of people who developed quantum mechanics; some are included to broaden the student's view of quantum mechanics, and some introduce non-conventional ways of viewing things. The asides are not essential reading, but the student will lose much if she ignores them; physics is, after all, a human pursuit.

I hope the reader will find this book relatively easy going. Chapters upon philosophical considerations and the history of quantum mechanics have been interspersed with the more technical chapters. Much which has been said before is repeated at appropriate points in the texts, and your author does not apologize for this but hopes that it makes the study of this text easier. Much that will follow in detail later has been lightly introduced in earlier chapters. There are concepts in quantum mechanics that, at first sight, hit the student like a brick wall; a gentle introduction is often less challenging. Even the brightest reader will not absorb the whole of this text in one reading, and perhaps three readings would be preferred. I hope the reader will find this book sufficiently clear and enlightening that the student will want to read it a second and third time.

Dennis Morris
May 2016

1

THE PLACE OF QUANTUM MECHANICS IN MODERN PHYSICS

Is the reader sitting comfortably? Then we will begin. Our world understanding is divided into several parts. There is the general theory of relativity, and there is the special theory of relativity. There is quantum field theory, which is usually referred to as QFT, and there is quantum mechanics. There is Newtonian mechanics, and there is string theory[1]. In this book, we are concerned with quantum mechanics but we will briefly review the other areas of physics and how they relate to quantum mechanics.

1.1   GENERAL RELATIVITY

General relativity is a theory that is distinct from and separate from quantum mechanics, and it is not part of our study of quantum mechanics. General relativity is a theory of gravity which includes within it special relativity. As such, general relativity is an essential part of humankind's understanding of the universe. Quantum mechanics is a theory that is not concerned with gravity and has

---

[1] We cannot really call string theory a physical theory because it is of no practical use.

nothing to say about gravity. It is of concern to theoretical physicists that our world understanding is divided into distinct and separate parts, and we would prefer to unite general relativity with quantum mechanics into only one theory, but, to date, humankind has failed in that endeavour. Part, possibly the whole, of the difficulty in uniting the theories of general relativity and quantum mechanics is that they are written in very different mathematical formulations. General relativity is written in the mathematics of tensors and covariant derivatives. Quantum mechanics is written in the mathematics of operators, eigenfunctions, eigenvectors, and eigenvalues (all of which are quite simple and will be explained shortly). Gravity is a necessarily continuous tensor field that can take any value. The dynamic variables of quantum mechanics are not tensors and are not necessarily continuous and they can take only allowed values.

Aside: There are other theories of gravity such as the Brans-Dicke theory. The Brans-Dicke theory is a scalar-tensor theory. This means that, as well as the tensor fields that we have in general relativity, we also have a scalar field that permeates the universe. The scalar field is the locally varying "strength" of the gravitational constant, G. Within general relativity, the gravitational constant is universal and does not vary from place to place. There is no evidence to decide between general relativity and the Brans-Dicke theory, but general relativity is preferred because it is the simpler of the two.

1.2 SPECIAL RELATIVITY

Special relativity is the mechanics of very rapidly moving bodies. We most often do not deal with very rapidly moving bodies in quantum mechanics, and so quantum mechanics is formulated as a non-relativistic theory. By this, we mean that we take the Newtonian view of energy as being:

$$E = K.E. + P.E. = \frac{p^2}{2m} + P.E. \qquad (1.1)$$

Rather than taking the relativistic view of energy as:

$$E^2 = p^2 c^2 + m^2 c^4 \tag{1.2}$$

The statement, "quantum mechanics is a non-relativistic theory" means no more and no less that this preference for the Newtonian expression for energy instead of the relativistic expression for energy.

Aside: What is energy? Does energy really exist or is it no more than a part of a set of mathematical relationships that describe mechanical processes? Do electrons really exist or are they no more than a part of a set of mathematical relationships, called quantum mechanics, that describe atomic processes? There are philosophical views of quantum mechanics that see electrons as no more than a part of a set of mathematical relationships. None-the-less, electric lights and vacuum cleaners do work.

Quantum mechanics is often said to be inaccurate and incomplete because it is not relativistic, but quantum mechanics does not exclude special relativity. Indeed, there are parts of quantum mechanics that certainly encompass special relativity, but only exceptionally do we actually need special relativity in quantum mechanics (very accurate calculation of spectra). Only in this sense, is quantum mechanics incomplete and inaccurate without special relativity. It is often said that, without special relativity, we would have no concept of intrinsic spin, but this too is not true. Intrinsic spin does naturally fall out of the relativistic Dirac equation, but intrinsic spin can also be deduced from no more than angular momentum commutation relations with no need of special relativity. There is nothing in special relativity that necessitates the concept of intrinsic spin, and there is nothing about intrinsic spin that necessitates special relativity.

Even though the theory of quantum mechanics is non-relativistic in its preference for the Newtonian energy expression, quantum mechanics is still a very useful theory because most day-to-day phenomena are non-relativistic. For these day-to-day phenomena, quantum mechanics gives the correct answers with very good precision.

As we said above, the mathematical formulation of quantum mechanics includes a type of vectors called eigenvectors. Special relativity is usually formulated in the mathematics of 4-vectors, otherwise known as Minkowski space-time, and there is a mathematical connection here but the connection is not patent within quantum mechanics. Only when we graduate to QFT does the connection become patent.

1.3   QUANTUM FIELD THEORY

Quantum field theory, QFT, is the relativistic version of quantum mechanics that incorporates centrally within it both quantum mechanics and the special theory of relativity. QFT deals with both the high energy physics of very rapidly moving particles and the low energy physics of non-relativistic particles. Calculations within QFT are much harder than calculations within quantum mechanics – which is why we use quantum mechanics most of the time.

QFT includes the relativistic energy relation:

$$E = mc^2 \qquad (1.3)$$

This relation allows the creation and annihilation of particles from, or into, energy, and so QFT can deal with the creation and annihilation of particles. Being non-relativistic in its view of energy, quantum mechanics does not include this relation, and so quantum mechanics is unable to deal with the creation and annihilation of particles.

QFT is formulated in the mathematics of the calculus of variations, Lagrangian action, and the Euler equations of motion. Such mathematics can be used in quantum mechanics, but, most often, particularly in introductory texts, such mathematics is not used in quantum mechanics.

Aside: It is possible to formulate general relativity as a Lagrangian action, and the mathematician David Hilbert (1862-1943) did so[2], but this does not lead to a unification of general relativity and QFT.

[2] D. Hilbert, Die Grundlagen der Physik (1915)

1.4   NEWTONIAN MECHANICS

Newtonian mechanics is the mechanics of electrically neutral, low velocity, macroscopic objects in weak gravitational fields – macroscopic means bigger than an atom. Newtonian mechanics is deterministic. Newtonian mechanics is founded on a set of dynamic variables that include energy, momentum, position, time, angular momentum, and others. In Newtonian mechanics, these dynamic variables are just assumed (observed) to exist. Within Newtonian mechanics, there is a set of relations between these dynamic variables such as the energy and momentum relation:

$$E = \frac{p^2}{2m} \qquad (1.4)$$

wherein E is energy, p is momentum, and m is mass.

Newtonian mechanics is remarkably successful, and it is by far the most widely used theory of mechanics in the modern world. However, experiments have found that Newtonian mechanics does not properly describe physical systems consisting of very small bodies. To deal with such systems of microscopic bodies, we need quantum mechanics. So, we are swapping Newtonian mechanics for quantum mechanics when we deal with tiny objects rather than large objects. This is very similar to the way we swap Newtonian mechanics for special relativity mechanics when we deal with very rapidly moving objects or we swap Newtonian gravity for general relativity when we deal with very strong gravitational fields. Mathematically, Newtonian mechanics is formulated as continuous relations between continuous real variables.

Newtonian mechanics does describe the slow moving macroscopic world accurately, and special relativity "morphs" into Newtonian mechanics at low velocity. So it is that quantum mechanics "morphs" into Newtonian mechanics for macroscopic systems. Quantum mechanics and Newtonian mechanics fit together seamlessly at the interface of the microscopic and the macroscopic and it might be said that quantum mechanics subcludes Newtonian mechanics in a way similar to the way that special relativity subcludes Newtonian mechanics.

1.5   STRING THEORY

String theory is a mathematically complex attempt to unify general relativity with QFT, and hence with quantum mechanics since QFT includes quantum mechanics. String theory is about vibrating strings set in a 10-dimensional space in which six of the dimensions are "compact" (rolled up). There are no practical uses, to date, of string theory, and its only physical prediction is the existence of a spin-2 boson assumed to be associated with the gravitational force. There is no observational support for string theory, and so it is considered to be speculative. String theory is not considered in this book.

1.6   QUANTUM MECHANICS

Quantum mechanics is the only mechanics that describes the behavior and properties of very small objects, like electrons or atoms. Quantum mechanics is often seen as the mechanics of objects that are sufficiently small and of sufficiently low energy that the act of observing one of them by hitting it with a photon[3] of light disturbs the object that is being observed – alters its momentum or position or energy or …. In the macroscopic world, it is presumed that hitting an object, say a cricket ball, with a photon of light to observe it will not noticeably disturb the object. It is this disturbance or non-disturbance by observation that is seen as differentiating the very small objects of quantum mechanics from the macroscopic objects of Newtonian mechanics. This disturbance is often associated with uncertainty in the values of some of the dynamic variables, but it is more accurate to associate quantum mechanical uncertainty with non-commuting operators having different eigenfunctions – which will be explained later – and to view quantum mechanics as the mechanics of dynamic operators with no mention of observational disturbance.

---

[3] The name photon was coined by G.N. Lewis in an article in Nature 18th December 1926.

Uncertainty is a central feature of quantum mechanics, but it does not exist in Newtonian mechanics. Within Newtonian dynamics, the values of all dynamic variables like position and momentum can, theoretically, be simultaneously known with infinite precision. Within quantum mechanics, the values of all dynamic variables like position and momentum cannot be simultaneously known with infinite precision.

Quantum mechanics is formulated as mathematical operators, eigenfunctions (eigenvectors), and eigenvalues. In quantum mechanics, the continuous dynamic variables, such as energy or momentum, of Newtonian mechanics are each taken to be associated with an operator. In quantum mechanics, for each dynamic Newtonian variable, there is one, and only one, corresponding operator such as the energy operator or the momentum operator. In Newtonian mechanics, the dynamic variables can take any (real) value. In quantum mechanics, the only allowed values of the dynamic variables, such as the energy or the momentum, are the eigenvalues of the associated operator. These eigenvalues are often discreet rather than continuous. Discreteness of the values of dynamical variables is a central feature of quantum mechanics. It is entirely to do with a discreet number of full wavelengths fitting into a given length as do standing waves in a vibrating string of given length.

In addition to operators corresponding to every Newtonian dynamic variable, quantum mechanics has other operators such as the intrinsic spin operators. There is no concept of intrinsic spin within Newtonian mechanics.

Operators are also different from Newtonian variables in that they often do not commute with each other, as we will explain later. There is no concept of non-commutation in Newtonian mechanics.

Newtonian mechanics is formulated in the geometric space, $\mathbb{R}^3$ together with an independent real time variable, $t \in \mathbb{R}$. Quantum mechanics is formulated in the geometric spaces formed by fitting together copies of the complex numbers, $\mathbb{C}$, at "right-angles" to each other, like $\mathbb{C}^2$ or $\mathbb{C}^3$, together with an independent real time variable, $t \in \mathbb{R}$. Such $\mathbb{C}^n$ spaces are called unitary spaces; the $\mathbb{R}^n$ spaces are called orthogonal spaces.

A vector in orthogonal (Newtonian) space, $\mathbb{R}^n$, is an ordered set of n real numbers. A vector in unitary (quantum mechanics) space, $\mathbb{C}^n$, is an ordered set of n complex numbers. We have:

$$\begin{bmatrix} 1 \\ 4 \\ 2 \end{bmatrix} \qquad : \qquad \begin{bmatrix} 1+2i \\ 3-7i \\ 2+3i \end{bmatrix} \qquad (1.5)$$

A vector in Orthogonal space : A vector in Unitary space

Newtonian Mechanics : Quantum Mechanics

This is a central difference between the two types of mechanics. The reader should note that time is added into $\mathbb{C}^n$ as a real variable rather than as a complex variable, which rather spoils the pattern.

Aside: The reader should be aware that there are other understandings of the nature of empty space. For example, quaternion space is a 4-dimensional space with one real axis and three imaginary axes. The Euclidean complex plane is a 2-dimensional space with one real axis and one imaginary axis. 2-dimensional space-time is the hyperbolic complex "plane" with one real axis and one imaginary axis.

The use of operator mathematics and of unitary space has led to quantum mechanics being seen as a mathematically daunting subject, but this is only because previous education has left us unfamiliar with operators and our hearts instinctively reject the "weird" mathematical operator formulation and the "weird" unitary space of quantum mechanics. If we can overcome our objection to the mathematical formulation of quantum mechanics, the subject is actually quite easy, as we hope the reader will discover. Easy or otherwise, an understanding of quantum mechanics is an essential part of our world understanding and is essential to modern engineering.

Quantum mechanics has become an everyday tool of engineers. It is used to design lasers, optic cables, transistors and microchips (think tunnel diode[4]) amongst many other now commonplace objects. It is the basis of magnetic resonance imaging (MRI scans),

---

[4] The tunnel diode is also known as Esaki diode. In 1973, Leo Esaki jointly won the physics Nobel prize for discovering the electron tunneling means by which these diodes work.

the scanning tunneling microscope, fluorescent light bulbs, ultra-precise thermometers, and ultra-precise atomic clocks. It is used in research as we try to build energy harvesters, try to build quantum computers, try to develop quantum cryptography, try to develop instantaneous communication, and try to develop teleportation. It is even used by biochemists as they seek to develop genetic engineering techniques. In spite of our instinctive dislike of its mathematical formulation, quantum mechanics does work marvellously well, and that is why it has a place of high regard in modern physics.

SUMMARY

To learn quantum mechanics, we are going to have to become familiar with the mathematics of operators, eigenvectors (eigenfunctions), and eigenvalues. We will also have to become familiar with the linear spaces in which these mathematical objects exist. We will have to become familiar with complex vectors, $\mathbb{C}^n$, and the inner products of those vectors. We will not have to become familiar with general relativity, special relativity, or string theory.

THE DYNAMIC VARIABLES IN QUANTUM MECHANICS

In any physical theory, there are particular dynamic variables such as energy, momentum, angular momentum etc. The energy variable appears in relativistic mechanics, in Newtonian mechanics, and in quantum mechanics. It is believed that energy will appear in every possible physical theory that describes the universe. There is a dynamic variable within quantum mechanics called intrinsic spin. Intrinsic spin does not appear as a dynamic variable within Newtonian mechanics or within relativistic mechanics, and so we see that different physical theories have different sets of dynamic variables. Different physical theories might also view the dynamic variables differently. In Newtonian mechanics, energy is just energy. In relativistic mechanics, energy is momentum in the time direction (or momentum is energy in the space direction). Different theories might relate variables to each other in different ways, and so we see that different theories that have the same set of dynamic variables do not necessarily treat these dynamic variables in the same way – that is part of being a different theory.

We would like an exact list of all the dynamic variables there are in the universe; we would like to know why there are that number and why they exist and how they relate to each other. Unfortunately, we do not yet properly understand these matters, but we might have some idea. That idea is called symmetry.

A physical system like a hydrogen atom has exactly the same properties regardless of whether it is pointing north or pointing west. Rotating experimental apparatus, say a test tube, by 90° does not affect the outcome of the experiment, say chemical reaction. We call this rotational symmetry. By using mathematics that is beyond this book (Noether's theorem), we can associate this symmetry with a dynamic variable. For rotational symmetry, that dynamic variable is angular momentum. The symmetry also leads to a conservation law. For rotational symmetry, that conservation law is conservation of angular momentum.

Experimental apparatus gives the same results if it is translated in space from one place to another. Equivalently, a hydrogen atom has the same properties in different places. We call this spatial translational symmetry; it is associated with conservation of linear momentum and hence with linear momentum. Experimental apparatus gives the same results if it is translated in time from one day to another. Equivalently, a hydrogen atom has the same properties on different days of the week. We call this temporal translational symmetry; it is associated with conservation of energy.

The reader might now think that all we have to do is write down the list of symmetries in the universe and we will have a list of dynamic variables. We believe that the reader would be correct, but we are unable to write down the list of symmetries in the universe. In particular, we do not understand the symmetry that, we assume, is associated with the intrinsic spin that we find in quantum mechanics[1].

Any physical theory is set over a type of geometric space. Special relativity is set over a hyperbolic 4-dimensional unified space-time. Newtonian mechanics is set in a Euclidean 3-dimensional space with an independent 1-dimensional time. Within these different types of space there are different types of symmetry. Within Newtonian space, we can rotate in a 2-dimensional plane as described by a single 2 × 2 rotation matrix but we cannot rotate 3-dimensionally

---

[1] Intrinsic spin seems to be associated with rotation through 720° rather than through 360°.

as described by a single 3×3 rotation matrix. In relativistic mechanics, we can rotate, 2-dimensionally, in space-time. Such rotation is a change of velocity. Rotation in space-time is not possible within Newtonian space because time is a separate thing from space within the Newtonian view. There are spaces, quaternion space is an example, in which we can do 4-dimensional rotations as described by a single 4×4 rotation matrix, and there are spaces, the C_3 spaces[2], in which we can do 3-dimensional rotations as described by a single 3×3 rotation matrix. An important symmetry in quantum mechanics is the 2-way oddness and evenness symmetry that we call parity. In the 3-dimensional C_3 spaces, parity is a 3-way "oddness and evenness"; in the 4-dimensional C_4 spaces, parity is a 4-way "oddness and evenness". In short, we will not understand the symmetries of the universe until we understand the space-time(s) of the universe. It is because we do not properly understand empty space that we cannot deduce a "correct" list of dynamic variables from symmetry alone.

There we have it. Your author would have liked to have started this book with something as simple as the real numbers, deduced the space-time(s) of the universe, taken the symmetries of that (those) space-time(s) and used them to deduce the existence of a particular set of dynamic variables which he would present in a tidy package to the reader. Your author, and, as far as he is aware, everyone else, is unable to do that because we do not understand empty space. None-the-less, the idea is good. If anyone asks the reader why we have angular momentum in quantum mechanics or why we have intrinsic spin in quantum mechanics or why we have energy in quantum mechanics, the reader will answer that they each are associated with a particular symmetry of empty space, we think. The mathematical expression of this idea is called Noether's theorem.

Of course, this list of dynamic variables would then be the list that every physical theory had to use and so we would be able to explain why there is energy in Newtonian mechanics and why there is momentum in relativistic mechanics.

[2.] See Dennis Morris: Complex Numbers – The Higher Dimensional Forms : ISBN: 978-0-955600-30-2

2.1   THE CORRESPONDENCE PRINCIPLE

Since we are not able to deduce the list of dynamic variables used in quantum mechanics "properly" from symmetries, we will have to use "stone-age" methods. Newtonian mechanics works well for macroscopic objects. Quantum mechanics ought to work at least as well for macroscopic objects, and it ought to give the same correct results. We therefore write down the list of Newtonian dynamic variables and we affirm that, for every Newtonian dynamic variable, there is a corresponding quantum mechanical dynamic variable. We further affirm that the relations between different Newtonian dynamic variables are duplicated between the corresponding quantum mechanical dynamic variables. These two affirmations are called the correspondence principle. It is an assertion and not a fact. The correspondence principle does not lead to a complete list of quantum mechanics dynamic variables; there is no intrinsic spin dynamic variable in Newtonian mechanics, but there is an intrinsic spin dynamic variable in quantum mechanics. However, for every Newtonian dynamic variable (like energy or momentum) there is a corresponding quantum mechanical dynamic variable.

The Correspondence Principle

For every Newtonian dynamic variable, there is a corresponding variable in quantum mechanics. The relations between the Newtonian dynamic variables are duplicated between the quantum mechanics variables.

The correspondence principle says that, since there is an energy variable in Newtonian mechanics, there is an energy variable in quantum mechanics, and, since there is a momentum variable in Newtonian mechanics, there is a momentum variable in quantum mechanics, and, since in Newtonian mechanics we have the kinetic energy:

$$E = \frac{1}{2}mv^2 = \frac{p^2}{2m} \qquad (2.1)$$

then in quantum mechanics, we have:

$$\hat{E} = \frac{\hat{p}^2}{2m} \tag{2.2}$$

Wherein we have put a carat (little hat) over the quantum mechanical variables to distinguish them from the Newtonian variables. It is because we use this non-relativistic expression for energy in quantum mechanics that quantum mechanics is considered to be a non-relativistic theory.

Aside: The fine structure constant is a constant that measures the strength of the electromagnetic force. It was first introduced by Arnold Sommerfield in 1916. It is defined as:

$$\alpha = \frac{e^2}{\hbar c} \tag{2.3}$$

Wherein e is the charge of the electron and c is the velocity of light. It is because the velocity of light is so large compared to everyday experience that we can use the non-relativistic form of energy in quantum mechanics. Alternatively, it is because the fine structure constant is so small that we can use the non-relativistic form of energy in quantum mechanics.

2.2 NEWTONIAN DYNAMIC VARIABLES CORRESPOND TO OPERATORS

We should warn the reader that, within quantum mechanics, the dynamic variables are presented as being operators that operate on a physical state expressed as a vector or as a wave function to extract a real number from that physical state. We will look at this in much more detail in later chapters. It will be left to the reader to decide whether or not the conventional operator view of dynamic variables is necessary or whether we can continue to look at the dynamic variables as being just variables. If we had been able to deduce a list of dynamic variables from symmetry considerations, we would not have been led to operators.

It is normal within quantum mechanics to denote an operator by putting a carat over it as we have done with the quantum mechanical variables (operators) above in (2.2).

There is more to the correspondence principle than immediately meets the eye. As we will see in later chapters, the order in which two different operators are applied matters. Operators are not necessarily commutative, and so, when we convert Newtonian variables to operators, we impose the commutation relations of the operators on to the physical system. Newtonian variables always commute with each other, and so we are assuming substantial changes to the physical system when we apply the correspondence principle. This imposition of commutation relations on to a physical system is called quantitisation of the physical system.

Operators are not the only mathematical objects that are associated with non-commutativity. There are non-commutative division algebras (higher dimensional types of complex numbers) like the quaternions and the A_3 algebras that are not commutative. Unfortunately, except for the quaternions (Clifford algebras are not division algebras), these non-commutative division algebras were not known a century ago when physicists discovered the need for non-commutativity in atomic physics. Thus, physicists turned to operators in their need for non-commutativity when perhaps they should have turned to non-commutative division algebras. As your author writes, there is research being done into this "other way of writing quantum mechanics", but, although there has been much success with electromagnetism, the research is not yet complete, and we do not yet know whether we can replace operators with non-commutative division algebras. If the reader thinks that operators are "unnatural" objects that ought not to be in a physical theory, the reader is not alone, but, to date, we cannot manage without them. Perhaps the reader will use the higher dimensional division algebras to rewrite quantum theory.

EXERCISE

1. The z-component of Newtonian angular momentum is given by:

$$L_z = x.p_y - y.p_x \qquad (2.4)$$

We see that the expression involves the Newtonian position variables $\{x, y\}$ and the Newtonian momentum variables $\{p_x, p_y\}$. Use the correspondence principle to find the form of the quantum mechanical z-angular momentum operator, $\hat{L}_z$?

CHAPTER 3

WAVE PARTICLE DUALITY, SUPERPOSITION, AND NON-LOCALITY

3.1   WAVE-PARTICLE DUALITY

In the early part of the 20th century, it was observed that particles like electrons have properties that are normally associated with waves and that electromagnetic waves, like light, have properties that are normally associated with particles. It has been, eventually and reluctantly, concluded by physicists that all objects are both waves and particles. We call this phenomenon wave-particle duality. Wave-particle duality is the observed fact which drove the development of quantum mechanics and which both justifies and insists upon a quantum mechanical formulation of our world understanding.

The wave-particle dual nature of objects leads to the breakdown of determinism and the use of probability in physics for which quantum mechanics is so famous. It is the dual nature of objects that leads to the de Broglie wave equation and to the Schrödinger wave equation of quantum mechanics. It is the dual nature of objects that leads to the Pauli exclusion principle, and to the quantitisation of what Newton saw as continuous dynamic variables.

This wave-particle duality is a duality. This is not a monality of existence, and it is not a triality of existence, or a quadrality of existence. Why duality? Why not triality? Why do objects not have wave properties and particle properties and, say, "water" properties thereby giving a triality of existence? Why do objects not have only particle properties? Why do waves exist at all?

To date, no explanation of the dual wave-particle nature of physical objects has been proposed. We observe the dual wave-particle nature of physical objects to be the case; experiments confirm the dual wave-particle nature of physical objects to be the case, but no-one has proffered an explanation of why the dual wave-particle nature of physical objects is the case.

3.2   SUPERPOSITION

One of the most beguiling facets of quantum mechanics is the superposition of states. In quantum mechanics, we assume that an unobserved system is not in a particular state but is in all possible states at the same time. Erwin Schrödinger, the man after whom the Schrödinger equation is named, presented this view with a dilemma about a cat.

3.3   SCHRÖDINGER'S CAT

The superposition of states was described by Schrödinger with the concept of a cat in a sealed box. There is a cat inside a sealed box. With the cat is a sealed bottle of poisonous vapor arranged such that the bottle will break if a radioactive nucleus decays. The decay or non-decay of a radioactive nucleus is a random event governed by chance. If the radioactive nucleus decays, the cat will be dead when an observer opens the box. If the radioactive nucleus does not decay, the cat will be alive when the observer opens the box. The state of being an alive cat corresponds to one state of the cat, and the state of being a dead cat corresponds to another state of the cat. As far as we are concerned, before we open the box, the cat is in a superposition of being both alive and dead. When we open the box, this superposition collapses into a single state; the cat is either dead or alive, but not both.

A physical system, be it a cat in a box or an electron in an atom, is seen in quantum mechanics as being in a superposition of possible states until it is observed. When the physical system is observed, the physical system collapses into only one of the possible states. This view is known as the Copenhagen interpretation of quantum mechanics.

3.4   THE COPENHAGEN INTERPRETATION OF QUANTUM MECHANICS

Between 1920 and 1924, Niels Bohr and Werner Heisenberg developed what has become known as the Copenhagen interpretation of quantum mechanics. The Copenhagen interpretation can be summed up as:

1. A physical system is a wave that is a linear sum of basis solutions (standing waves) of the Schrödinger wave equation (the linear sum is called a wave function). We see this linear sum of basis solutions as a superposition of all possible solutions. This superposition of basis solutions evolves both smoothly and deterministically in time until a measurement of some property of the system is made at which point the superposition probabilistically (not deterministically) collapses into a particular basis solution called an eigenstate.

2. Nature is not deterministic.

3. It is not possible to simultaneously know the value of every property of the physical system.

4. Everything exists in a dual wave/particle form.

This is not the only interpretation of quantum mechanics. Another interpretation is the decoherence interpretation founded by David Bohm (1917–1992) and later used by Hugh Everett (1930–1982) in the many worlds interpretation.

Aside: The German physicist Werner Karl Heisenberg (1901–1976) is famous for his formulation of matrix mechanics circa 1925[1] and for his proposal of the uncertainty principle in 1927. He was awarded the physics Nobel prize in 1932. Although Heisenberg first proposed the uncertainty principle, the principle was deduced more formally by E. Kennard[2] (1885–1968) in 1927 and by Hermann Weyl[3] (1885–1955) in 1928.

There is much dispute over Heisenberg's role in the second world war. He certainly worked upon developing the atomic bomb for the Nazis, but he also failed to develop the atomic bomb for the Nazis. His friends say he deliberately misled the atomic bomb research. Others say that his German patriotism motivated him to collaborate with the Nazis and that he failed to build the atomic bomb in spite of all his efforts.

3.5 NON-LOCALITY

There are phenomena within quantum mechanics that are associated with instantaneous communication from one locality in space to another locality in space. Objects seem to be able to communicate instantaneously from place to place. Such phenomena are said to be non-local, and they have been observed in experiments.

Aside: Within physics, we encounter two types of trigonometric functions. The hyperbolic trigonometric functions {cosh( ), sinh( )} are the trigonometric functions of the hyperbolic complex plane, $\mathbb{S}$,

---

[1] W. Heisenberg. Z f Physik 33, 879 (1925).
[2] Kennard. E.H. (1927) - Zur Quantenmechanik einfacher Bewegungstype. Zeitschrift für Physik 44, 4–5, 326.
[3] Weyl H. Gruppentheorie und Quantenmechanik. Leipzig Hiizel.

and are associated with one time dimension and one space dimension. Particles move in only one spatial direction. The Euclidean trigonometric functions {cos( ), sin( )} are the trigonometric functions of the complex plane, $\mathbb{C}$, and are not associated with time but are associated with two space dimensions. Waves move in all spatial directions. The hyperbolic trigonometric functions are associated with symmetric matrix variables. The Euclidean trigonometric functions are associated with anti-symmetric matrix variables. It seems that the mathematics of physics comes in two types, and we might expect some kind of dual nature in physics. Since the Euclidean trigonometric functions are associated with waves, we might expect half of this dual nature to be wave-like. Since the hyperbolic trigonometric functions are associated with a single spatial direction, we might expect half of this dual nature to be particle-like.

The mathematics of the Euclidean space $\mathbb{C}$ do not include time, and so we might expect instantaneous communication, which we call non-locality, within such space. If we use the mathematics of the complex numbers $\mathbb{C}$ to describe the possible states of a physical system, then, without time, the system must be in every one of these possible states at the same time - superposition. When an observer within time interferes with the superposition by forcing time into the physical system, the physical system will be in a particular state rather than all possible states. Withdraw the concept of time from a physical system, and it will be in all the states that it ever was in or ever could be in at the same time; this is superposition.

We leave it to the reader to evaluate the merits and demerits of this explanation.

3.6   MANY PATHS

Within quantum mechanics, there is uncertainty in the position and momentum of a particle. We will come to this in later chapters. Uncertainty is that we cannot simultaneously know both the exact momentum of a particle and the exact position of that particle. This means that particles do not have definite trajectories through space. We might observe a particle to be at position A one moment and to be at position B at a later moment, and so we know that the particle has traveled from A to B. However, we are unable to say which path the particle followed to get from A to B. This is not just a failing of our knowledge; this is within the very nature of particles and space and time.

There is an interpretation of quantum mechanics made known by Richard Feynman (1918–1988) in which particles are taken to travel from A to B by every possible path, including those that visit the distant edges of the known universe. The path is a superposition of all possible paths. Within this interpretation, quantum mechanics is done using path integrals which we do not cover within this book.

EXERCISES

1. A slowly vibrating string is bowed to the north on Tuesday and is straight on Wednesday and is bowed to the south on Thursday and is straight again on Friday and completes its cycle by being bowed to the north on Saturday – I did say the string was vibrating slowly. Take away time. In what state is the string?

2. A particle moves in a straight line through the complex plane, $\mathbb{C}$, from the point $a + ib$ to the point $c + id$. Is it at all points along its route at the same time?

3. Is being in two places at the same time instantaneous travel at infinite speed?

4. Calculate:

$$\exp\left(\begin{bmatrix} a & b \\ -b & a \end{bmatrix}\right) = \begin{bmatrix} e^a & 0 \\ 0 & e^a \end{bmatrix} \exp\left(\begin{bmatrix} 0 & b \\ -b & 0 \end{bmatrix}\right) \tag{3.1}$$

Note that:

$$\mathbb{C} = a + ib = \begin{bmatrix} a & b \\ -b & a \end{bmatrix} \tag{3.2}$$

Look at the graphs of the four trigonometric functions $\{\cosh(\ ), \sinh(\ )\}$ and $\{\cos(\ ), \sin(\ )\}$ (next chapter). Why are waves associated with the complex numbers $\mathbb{C}$?

CHAPTER 4

COMPLEX NUMBERS, WAVE EQUATIONS, AND THE MOMENTUM OPERATOR

A wave equation is an equation that has solutions that involve the Euclidean trigonometric functions {cos( ), sin( )}. These trigonometric functions have graphs that are waves, and so it is that equations that have these trigonometric functions as solutions are equations that describe waves.

The Cosine Function

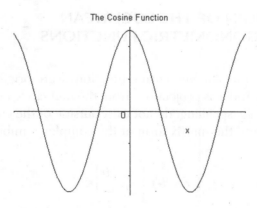

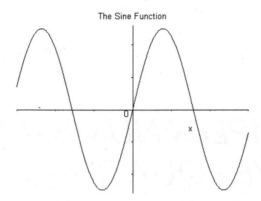

The Sine Function

The Euclidean trigonometric functions, as with all trigonometric functions, are such that they have a differentiation cycle:

$$\frac{\partial}{\partial x}(\sin x) = \cos x$$
$$\frac{\partial}{\partial x}(\cos x) = -\sin x \tag{4.1}$$

We therefore find them involved in solutions of differential equations like:

$$\frac{\partial^2 x}{\partial x^2} = -x \tag{4.2}$$

4.1   ORIGIN OF THE EUCLIDEAN TRIGONOMETRIC FUNCTIONS

Although the sine and cosine functions were originally defined in 6th century India as projections from the unit circle on to the axes, they, technically speaking, do not exist outside of the complex numbers, $\mathbb{C}$. We use the matrix form of the complex numbers for pedagogical ease:

$$a + ib \equiv \begin{bmatrix} a & b \\ -b & a \end{bmatrix} \in \mathbb{C} \tag{4.3}$$

The reader might like to multiply two such matrices together to convince herself that the above 2×2 matrix is the complex numbers.

We have:

$$\exp\left(\begin{bmatrix} a & b \\ -b & a \end{bmatrix}\right) = \begin{bmatrix} r & 0 \\ 0 & r \end{bmatrix}\begin{bmatrix} \cos b & \sin b \\ -\sin b & \cos b \end{bmatrix} \qquad (4.4)$$

wherein we see that the rotation matrix within the complex numbers, $\mathbb{C}$, contains the sine and cosine functions. Because these trigonometric functions are within the complex numbers, $\mathbb{C}$, all wave equations are associated with the complex numbers. This is not immediately apparent in the case of the Newtonian wave equation for an ideal string which we will meet shortly; however, when we deal with this wave equation, we will see, with thought, that it is connected to the complex numbers.

In contrast, we look at the hyperbolic trigonometric functions of the hyperbolic complex numbers, $\mathbb{S}$, that are the algebra of 2-dimensional space-time.

Aside: The reader should note that the whole of the theory of special relativity can be written within the hyperbolic complex numbers (and its 4-dimensional friend) and that doing so leads automatically to an expanding universe with an inflationary beginning and a changing rate of expansion. The presentation is also mathematically flawless, which is not true of the 4-vector presentation of special relativity (the acceleration 4-vector has to be fudged)[1].

We have:

$$\exp\left(\begin{bmatrix} a & b \\ b & a \end{bmatrix}\right) = \begin{bmatrix} h & 0 \\ 0 & h \end{bmatrix}\begin{bmatrix} \cosh b & \sinh b \\ \sinh b & \cosh b \end{bmatrix} \qquad (4.5)$$

These trigonometric functions have nothing to do with waves.

[1.] See: Empty Space is Amazing Stuff by Dennis Morris: Pantaneto Press ISBN: 978-0-9549780-7-5

The Cosh and the Sinh Trigonometric Functions

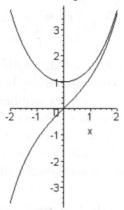

There is a thing about the Euclidean trigonometric functions, {cos( ), sin( )}, that is of fundamental importance within quantum mechanics; they are associated with special numbers. If we look at the graphs above, we see that the sine and cosine functions repeat themselves every 2π, and so we have the special numbers {0, 2π, 4π,}. This repetition is ultimately due to the fact that we can rotate all the way through 360° within the complex plane. We have no such special numbers associated with the hyperbolic trigonometric functions because we cannot rotate all the way through 360° in the hyperbolic complex plane; we cannot get beyond the speed of light.

Perhaps more importantly, there are special numbers associated with fitting the Euclidean trigonometric functions, {cos( ), sin( )}, into a definite length. We have:

Sin(x/2) and Sin(x)

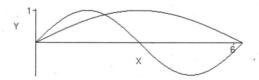

Sin(3x/2) and Sin(2x)

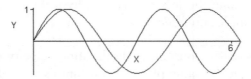

We see that to get a "complete" graph that starts at zero and returns to zero, we need to use the special numbers $\left\{\frac{1}{2},1,\frac{3}{2},2...\right\}$. There is nothing like this in hyperbolic trigonometry. We will come across exactly this again when we look at standing waves on an ideal Newtonian string. This is why there are special numbers (discreet energies) associated with electrons in orbit around an atomic nucleus. The electrons are orbital standing waves whose ends must fit together neatly.

4.2   SCALING PARAMETERS AND PHYSICAL CONSTANTS

In general, the 2-dimensional complex numbers, $\mathbb{C}$, and the 2-dimensional hyperbolic complex numbers, $\mathbb{S}$, are both of the form:

$$\exp\left(\begin{bmatrix} a & b \\ \lambda b & a \end{bmatrix}\right) : \lambda \neq 0 \qquad (4.6)$$

λ is a scaling parameter that measures the units used to measure along one axis against the units used to measure along the other axis. When $\lambda > 0$, we have the hyperbolic complex numbers which are space-time. The distance function of 2-dimensional space-time is $d^2 = t^2 - \frac{1}{c^2}z^2$ where c is the limiting velocity (the velocity of light). We have:

$$\det\left(\begin{bmatrix} t & z \\ \lambda z & t \end{bmatrix}\right) = t^2 - \lambda z^2 \qquad (4.7)$$

We see that $\frac{1}{\lambda} = c^2$, and we have a physical constant. Of course, by adjusting the units in which we measure space (or time), we can set $c = 1$, but, none-the-less, we have a physical constant within the hyperbolic complex numbers.

Aside: The physical constant is the limiting velocity within space-time. This is not the same physical constant as the velocity of light (electromagnetic waves); the physical constant is a space-time constant and not an electromagnetic constant. It is only coincidence, if there is such a thing in the universe, that light travels at the limiting velocity. We ought to use different symbols for the limiting velocity and the speed of light because they are different things.

When $\lambda < 0$, we have the Euclidean complex numbers, $\mathbb{C}_\lambda$. (We have subscripted a λ to avoid confusion with the $\lambda = 1$ form, $\mathbb{C}$.) These too contain a physical constant. For clarity, we will keep $\lambda > 0$ and put a minus sign before λ to take account of its negativity.

4.3   THE MOMENTUM OPERATOR

We will differentiate a complex function with respect to the imaginary axis.

$$
\frac{\partial \begin{bmatrix} f(a,b) & g(a,b) \\ -\lambda g(a,b) & f(a,b) \end{bmatrix}}{\partial \begin{bmatrix} 0 & b \\ -\lambda b & 0 \end{bmatrix}} = \frac{1}{\begin{bmatrix} 0 & 1 \\ -\lambda & 0 \end{bmatrix}} \frac{\partial \begin{bmatrix} f(a,b) & g(a,b) \\ -\lambda g(a,b) & f(a,b) \end{bmatrix}}{\partial \begin{bmatrix} b & 0 \\ 0 & b \end{bmatrix}}
$$

$$
= \begin{bmatrix} 0 & -\dfrac{1}{\lambda} \\ 1 & 0 \end{bmatrix} \begin{bmatrix} \dfrac{\partial f}{\partial b} & \dfrac{\partial g}{\partial b} \\ -\lambda \dfrac{\partial g}{\partial b} & \dfrac{\partial f}{\partial b} \end{bmatrix} \tag{4.8}
$$

$$
= \begin{bmatrix} 0 & -1 \\ \lambda & 0 \end{bmatrix} \begin{bmatrix} \dfrac{1}{\lambda} & 0 \\ 0 & \dfrac{1}{\lambda} \end{bmatrix} \begin{bmatrix} \dfrac{\partial f}{\partial b} & \dfrac{\partial g}{\partial b} \\ -\lambda \dfrac{\partial g}{\partial b} & \dfrac{\partial f}{\partial b} \end{bmatrix}
$$

We see that the process of differentiation by an imaginary variable necessitates multiplication by the negative of the imaginary unit, $-i_\lambda$ and by the inverse of the scaling parameter, $\frac{1}{\lambda}$. We might, in other notation, write this differentiation operation as:

$$-i\frac{1}{\lambda}\frac{\partial}{\partial b} \qquad (4.9)$$

Within quantum mechanics, we will soon meet the quantum mechanical momentum operator:

$$\widehat{p_x} = -i\hbar\frac{\partial}{\partial x} \qquad (4.10)$$

Wherein $\hbar = \frac{h}{2\pi}$ is known as "aitch bar". The constant h, known as Planck's constant, is a fundamental physical constant within quantum mechanics. We see that:

$$\hbar \equiv \frac{1}{\lambda} \qquad (4.11)$$

Of course, by choosing appropriate units, we can set $\hbar = 1$. The momentum operator within quantum mechanics is just the algebraic operation of differentiation with respect to the imaginary variable. This operator plays a central role in quantum mechanics. Aitch bar, $\hbar$, is a physical constant that also plays a central role in quantum mechanics.

Aside: The mass dimensions of $\hbar$ are ML^2T^{-2}. This is called action; it is equivalent to length × momentum or time × energy.

4.4 EIGENFUNCTIONS OF THE MOMENTUM OPERATOR

The rotation matrix of the scaled complex numbers, $\mathbb{C}_\lambda$, is:

$$\exp\left(\begin{bmatrix} 0 & b \\ -\lambda b & 0 \end{bmatrix}\right) = \begin{bmatrix} \cos\left(\sqrt{\lambda}b\right) & \frac{1}{\sqrt{\lambda}}\sin\left(\sqrt{\lambda}b\right) \\ -\lambda\frac{1}{\sqrt{\lambda}}\sin\left(\sqrt{\lambda}b\right) & \cos\left(\sqrt{\lambda}b\right) \end{bmatrix} \qquad (4.12)$$

Let us differentiate this rotation matrix of the $\mathbb{C}_\lambda$ algebra with respect to the imaginary variable. We will do it with an angle of $b = n\theta$:

$$\frac{\partial \begin{bmatrix} \cos(n\sqrt{\lambda}\theta) & \frac{1}{\sqrt{\lambda}}\sin(n\sqrt{\lambda}\theta) \\ -\lambda\frac{1}{\sqrt{\lambda}}\sin(n\sqrt{\lambda}\theta) & \cos(n\sqrt{\lambda}\theta) \end{bmatrix}}{\partial \begin{bmatrix} 0 & \theta \\ -\lambda\theta & 0 \end{bmatrix}} \quad (4.13)$$

$$= \begin{bmatrix} 0 & -\frac{1}{\lambda} \\ 1 & 0 \end{bmatrix} \begin{bmatrix} \frac{\partial}{\partial\theta}\cos(n\sqrt{\lambda}\theta) & \frac{1}{\sqrt{\lambda}}\frac{\partial}{\partial\theta}\sin(n\sqrt{\lambda}\theta) \\ -\lambda\frac{1}{\sqrt{\lambda}}\frac{\partial}{\partial\theta}\sin(n\sqrt{\lambda}\theta) & \frac{\partial}{\partial\theta}\cos(n\sqrt{\lambda}\theta) \end{bmatrix}$$

$$= \begin{bmatrix} 0 & -1 \\ \lambda & 0 \end{bmatrix} \begin{bmatrix} \frac{1}{\lambda} & 0 \\ 0 & \frac{1}{\lambda} \end{bmatrix} \begin{bmatrix} -n\sqrt{\lambda}\sin(n\sqrt{\lambda}\theta) & \frac{n\sqrt{\lambda}}{\sqrt{\lambda}}\cos(n\sqrt{\lambda}\theta) \\ -\lambda\frac{n\sqrt{\lambda}}{\sqrt{\lambda}}\cos(n\sqrt{\lambda}\theta) & -n\sqrt{\lambda}\sin(n\sqrt{\lambda}\theta) \end{bmatrix}$$

$$\quad (4.14)$$

$$= \begin{bmatrix} n & 0 \\ 0 & n \end{bmatrix} \begin{bmatrix} \cos(n\sqrt{\lambda}\theta) & \frac{1}{\sqrt{\lambda}}\sin(n\sqrt{\lambda}\theta) \\ -\lambda\frac{1}{\sqrt{\lambda}}\sin(n\sqrt{\lambda}\theta) & \cos(n\sqrt{\lambda}\theta) \end{bmatrix}$$

We see that differentiation with respect to the imaginary variable of the rotation matrix with angle $n\theta$ simply multiplies the rotation matrix by n. We say that the rotation matrix with angle $n\theta$ is an eigenfunction (special function) of the differentiate with respect to the imaginary variable operator. We say that the number, n, is the eigenvalue (special value) associated with that particular eigenfunction. Notice that each value of n corresponds to a single eigenfunction.

If, instead of $n\theta$, we had used the angle $n\lambda\theta$, the rotation matrix would be multiplied by $n\lambda$, and, using other notation, we could have written this as:

$$-i_\lambda \frac{1}{\lambda}\frac{\partial}{\partial\theta}\left(e^{i_\lambda n\lambda\theta}\right) = -i_\lambda \frac{1}{\lambda}i_\lambda n\lambda e^{i_\lambda n\lambda\theta} = ne^{i_\lambda n\lambda\theta} \quad (4.15)$$

This is equivalent to the momentum operator:

$$-i\hbar \widehat{\frac{\partial}{\partial x}}\left(e^{i\frac{n}{\hbar}x}\right) = ne^{i\frac{n}{\hbar}x} \tag{4.16}$$

The reader will become familiar with this expression in later chapters. The value, n, is taken to be the linear momentum in the x-direction. The eigenfunctions $e^{i\frac{p_x}{\hbar}x}$ are said to be the x-momentum eigenfunctions.

4.5   THE ZERO POTENTIAL ENERGY OPERATOR

Let us differentiate twice with respect to the imaginary variable. We have:

$$
\begin{bmatrix} 0 & -1 \\ \lambda & 0 \end{bmatrix}
\begin{bmatrix} \frac{1}{\lambda} & 0 \\ 0 & \frac{1}{\lambda} \end{bmatrix}
\frac{\partial}{\partial}
\frac{\begin{bmatrix} \frac{\partial f}{\partial b} & \frac{\partial g}{\partial b} \\ -\lambda\frac{\partial g}{\partial b} & \frac{\partial f}{\partial b} \end{bmatrix}}{\begin{bmatrix} 0 & -b \\ \lambda b & 0 \end{bmatrix}}
$$

$$
= \begin{bmatrix} 0 & -1 \\ \lambda & 0 \end{bmatrix}
\begin{bmatrix} \frac{1}{\lambda} & 0 \\ 0 & \frac{1}{\lambda} \end{bmatrix}
\begin{bmatrix} 0 & -\frac{1}{\lambda} \\ 1 & 0 \end{bmatrix}
\begin{bmatrix} \frac{\partial^2 f}{\partial b^2} & \frac{\partial^2 g}{\partial b^2} \\ -\lambda\frac{\partial^2 g}{\partial b^2} & \frac{\partial f}{\partial b^2} \end{bmatrix} \tag{4.17}
$$

$$
= \begin{bmatrix} -\lambda & 0 \\ 0 & -\lambda \end{bmatrix}
\begin{bmatrix} \frac{1}{\lambda^2} & 0 \\ 0 & \frac{1}{\lambda^2} \end{bmatrix}
\begin{bmatrix} \frac{\partial^2 f}{\partial b^2} & \frac{\partial^2 g}{\partial b^2} \\ -\lambda\frac{\partial^2 g}{\partial b^2} & \frac{\partial f}{\partial b^2} \end{bmatrix}
$$

We can write this as:

$$-\lambda\hbar^2\frac{\partial^2\psi}{\partial b^2} \quad : \quad \psi \in \mathbb{C}_\lambda \tag{4.18}$$

In due course, we will meet the quantum mechanical zero potential energy operator, also called the zero potential Hamiltonian (or zero potential Hamiltonian operator). There are two ways of denoting the zero potential Hamiltonian $\{\widehat{H}, \widehat{E}\}$. That operator is:

$$\widehat{H} = \widehat{E} = -\frac{\hbar^2}{2m}\frac{\partial \psi}{\partial x^2} \tag{4.19}$$

Aside: The Hamiltonian is named after the Irish mathematician and physicist William Rowan Hamilton (1805–1865). Hamilton reformulated Newtonian mechanics to produce Hamiltonian mechanics. Hamiltonian mechanics is central to much modern physics including quantum mechanics.

In 1843, Hamilton also discovered the 4-dimensional quaternion division algebra. He found this after mathematicians had spent, without success, some 200 years searching for division algebras of dimension higher that two. His discovery was followed in 1848 by Cockle's discovery of the 2-dimensional hyperbolic complex numbers[2].

Within quantum mechanics, the Hamiltonian operator is derived from the correspondence principle and the non-relativistic kinetic energy relation:

$$E = \frac{p^2}{2m}$$

$$\widehat{E} = \frac{\widehat{p_x}^2}{2m} = \frac{1}{2m}\left(-i\hbar\frac{\partial}{\partial x}\right)^2 = -\frac{\hbar^2}{2m}\frac{\partial^2}{\partial x^2} \tag{4.20}$$

We see there is a difference in that $\lambda \leftrightarrow \frac{1}{2m}$. None-the-less, we see a striking similarity between double differentiation with respect to the imaginary variable and the energy operator.

4.6   ENERGY EIGENFUNCTIONS

Let us double differentiate this rotation matrix of the $\mathbb{C}\lambda$ algebra with respect to the imaginary variable. We will do it with an angle of $b = n\theta$:

---

[2] Cockle. On the symbols of algebra, and the theory of tessarines. Phil. Mag. (3), 34, 406-410

$$\frac{\partial^2 \begin{bmatrix} \cos(n\sqrt{\lambda}\theta) & \frac{1}{\sqrt{\lambda}}\sin(n\sqrt{\lambda}\theta) \\ -\lambda\frac{1}{\sqrt{\lambda}}\sin(n\sqrt{\lambda}\theta) & \cos(n\sqrt{\lambda}\theta) \end{bmatrix}}{\partial \begin{bmatrix} 0 & \theta \\ -\lambda\theta & 0 \end{bmatrix}^2}$$

$$= \begin{bmatrix} n & 0 \\ 0 & n \end{bmatrix} \frac{\partial \begin{bmatrix} \cos(n\sqrt{\lambda}\theta) & \frac{1}{\sqrt{\lambda}}\sin(n\sqrt{\lambda}\theta) \\ -\lambda\frac{1}{\sqrt{\lambda}}\sin(n\sqrt{\lambda}\theta) & \cos(n\sqrt{\lambda}\theta) \end{bmatrix}}{\partial \begin{bmatrix} 0 & \theta \\ -\lambda\theta & 0 \end{bmatrix}} \quad (4.21)$$

$$= \begin{bmatrix} n^2 & 0 \\ 0 & n^2 \end{bmatrix} \begin{bmatrix} \cos(n\sqrt{\lambda}\theta) & \frac{1}{\sqrt{\lambda}}\sin(n\sqrt{\lambda}\theta) \\ -\lambda\frac{1}{\sqrt{\lambda}}\sin(n\sqrt{\lambda}\theta) & \cos(n\sqrt{\lambda}\theta) \end{bmatrix}$$

We see that the rotation matrix, as well as being an eigenfunction of the momentum operator, is also an eigenfunction of the zero potential energy operator but with eigenvalues n^2 rather than n. The reader should be aware that this is the case only where we have a zero potential.

If, instead of $n\theta$, we had used the angle $\lambda n\sqrt{2m}\theta$, the rotation matrix would be multiplied by $2m\lambda^2 n^2$, and this is equivalent to:

$$-\frac{\hbar^2}{2m}\frac{\partial^2}{\partial x^2}\left(e^{i\frac{\sqrt{2Em}}{\hbar}x}\right) = Ee^{i\frac{\sqrt{2Em}}{\hbar}x} \quad (4.22)$$

The E is taken to be the energy.

SUMMARY

There are special numbers (eigenvalues) associated with waves. Waves are associated with the complex number division algebra $\mathbb{C}$ through the trigonometric functions of that algebra. The scaling parameters of division algebras seem to be associated with physical constants. The everyday algebraic operation of differentiation is portrayed within quantum mechanics as an operator. Operators have special functions (eigenfunctions) and special numbers (eigenvalues) associated with them.

We will meet the momentum operator much within the rest of this book. We have seen above that it is no more than the operation of differentiation with respect to the imaginary axis within the complex numbers, $\mathbb{C}_\lambda$. Most books on quantum mechanics do not present it in this way. We could have presented the momentum operator with $\lambda = 1$, but we would not have found a physical constant and would not have had to feed $\hbar$ in by hand. We will speak much of operators in later chapters. Quantum mechanics is formulated in terms of operators, but do we really need them? Perhaps operators are no more than normal algebraic operations, and quantum mechanics might be more tidily formulated without them.

One intriguing thing about the above is the association between momentum (and energy) and rotation in the complex plane. It is fair to say that we do not properly understand why there should be such an association.

Aside: We point out the conjugate of e^{ix} is e^{-ix} and that, for a complex number, $\Psi = \Psi_{\text{Real}} + i\Psi_{\text{Imaginary}}$, we have the modulus squared is $|\psi|^2 = \psi\psi^*$.

We finish this chapter with an aside that is more to do with QFT than quantum mechanics. We put it in for the edification of the reader, and we use the standard $\lambda = 1$ form of the complex numbers.

Aside: Let us rotate the zero potential energy operator matrix:

$$\begin{bmatrix} -\dfrac{\partial^2 f}{\partial y^2} & -\dfrac{\partial^2 g}{\partial y^2} \\[2ex] \dfrac{\partial^2 g}{\partial y^2} & -\dfrac{\partial^2 f}{\partial y^2} \end{bmatrix} \begin{bmatrix} \cos\theta & \sin\theta \\ -\sin\theta & \cos\theta \end{bmatrix} =$$

$$\begin{bmatrix} -\dfrac{\partial^2 f}{\partial y^2}\cos\theta & -\dfrac{\partial^2 f}{\partial y^2}\sin\theta \\[2ex] \dfrac{\partial^2 f}{\partial y^2}\sin\theta & -\dfrac{\partial^2 f}{\partial y^2}\cos\theta \end{bmatrix} + \begin{bmatrix} \dfrac{\partial^2 g}{\partial y^2}\sin\theta & -\dfrac{\partial^2 g}{\partial y^2}\cos\theta \\[2ex] \dfrac{\partial^2 g}{\partial y^2}\cos\theta & \dfrac{\partial^2 g}{\partial y^2}\sin\theta \end{bmatrix}$$

$$(4.23)$$

If we now set $\theta = 0$ in these matrices, we get one Hamiltonian (energy) matrix out of phase with the other Hamiltonian. We get an operator of the form:

$$\widehat{H} = -\frac{\hbar^2}{2m}\frac{\partial^2}{\partial x^2} + V \qquad (4.24)$$

This is the non-zero potential quantum mechanical energy operator (the non-zero potential Hamiltonian). It seems that a potential is a thing to do with phase, and this is indeed the way it is seen in QFT.

EXERCISES

1. Calculate:

$$\begin{bmatrix} a & b \\ -\lambda b & a \end{bmatrix}\begin{bmatrix} c & d \\ -\lambda d & c \end{bmatrix} \qquad (4.25)$$

    Is the product of the same form as the two factors?

2. At what points does the function $f(x) = \cos(4x)$ cross the x-axis?

3. Using $d = e^a$, what is the determinant of:

$$\exp\left(\begin{bmatrix} a & b \\ b & a \end{bmatrix}\right) = \begin{bmatrix} e^a & 0 \\ 0 & e^a \end{bmatrix}\begin{bmatrix} \cosh b & \sinh b \\ \sinh b & \cosh b \end{bmatrix} \qquad (4.26)$$

    And what is the determinant of:

$$\begin{bmatrix} t & z \\ z & t \end{bmatrix} \qquad (4.27)$$

    Putting these two determinants equal, what do you have?

4. Using $d = e^a$, what is the determinant of:

$$\exp\left(\begin{bmatrix} a & b \\ -b & a \end{bmatrix}\right) = \begin{bmatrix} e^a & 0 \\ 0 & e^a \end{bmatrix}\begin{bmatrix} \cos b & \sin b \\ -\sin b & \cos b \end{bmatrix} \qquad (4.28)$$

And what is the determinant of:

$$\begin{bmatrix} x & y \\ -y & x \end{bmatrix} \tag{4.29}$$

Putting these two determinants equal, what do you have?

CHAPTER 5

AN END TO DETERMINISM

If a polarized beam of light meets a polarizer oriented parallel to the polarization of the light, all the light will pass through the polarizer. If a polarized beam of light meets a polarizer oriented perpendicularly to the polarization of the light, none the light will pass through the polarizer. If a polarized beam of light meets a polarizer oriented at 45° to the polarization of the light, exactly half the light will pass through the polarizer. Polarization happens because light is a wave. If the beam of light that meets a polarizer oriented at 45° to the polarization consists of only one photon of light, then, because photons cannot be divided into pieces, the photon either passes through the polarizer or it does not pass through the polarizer, but there is no way for the polarizer to decide which of these options will happen. All photons are the same, but sometimes a photon will pass through the 45° polarizer and sometimes a photon will not pass through the 45° polarizer. We have two identical events, which are essentially the same event, with different outcomes. Classical physics presumed that identical events would always produce identical outcomes. This presumption is called determinism. The fact that light comes in photons (wavy-particle things) means that the universe cannot be deterministic. Chance plays a part in the evolution of the universe.

It might be that there is the detonator of a nuclear bomb at the other side of the polarizer and that this detonator is such that, if a photon of light hits the detonator, the bomb explodes, nuclear war breaks out, and the Earth is destroyed. Whether or not the Earth is destroyed or survives depends upon whether or not the photon goes through the 45° polarizer. The destruction of the Earth is a matter of chance.

The reader might think that there must be some hidden variable within the photons that distinguishes one from another. Albert Einstein thought this. He expressed it in the famous phrase, "God does not play dice." It was shown by J. S. Bell (1928–1990) in 1964 that there is no such (local) hidden variable[1].

Bell's Theorem

No physical theory of local hidden variables can produce all the predictions of quantum mechanics.

Bell's theorem has been experimentally verified by Clauser and Shimony[2] in 1978 and by Alain Aspect et al in 1981[3] and 1982.[4]

Aside: In 2008, the university of Toronto inaugurated the "John Stewart Bell" prize for research into the fundamental issues of quantum mechanics. The first recipient in 2009 was Alain Aspect.

This "overthrow of determinism" is probably the most profound change brought to human understanding by quantum mechanics. Quantum mechanics is different from classical mechanics in many ways, but, by far, the most important difference is this change from the deterministic view of the world to the non-deterministic view of the world. It follows directly from the view that a light wave is a particle; it follows directly from wave-particle duality.

---

[1] J.S.Bell On the Einstein, Podolsky, Rosen Paradox. Physics, 1, 3, 195–200 (1964).
[2] J.F.Clauser & A.Shimony Bell's Theorem: experimental tests and implications. Reports on progress in Physics 41, 1881, (1978).
[3] Aspect et al. Experimental tests of realistic local theories via Bell's theorem. Phy Rev Lett 47, 460, (1981).
[4] A. Aspect, J. Dalibard, & C. Roger. Physical Review Letters Vol. 49 pgs 91 & 1804 : December 1982.

5.1   MANY UNIVERSES

Suppose that, when the polarized photon hits the 45° polarizer, the universe doubles into two universes. In one of these two universes, the photon does not pass through the polarizer and the Earth is not destroyed. In the other of these two universes, the photon does pass through the polarizer and the Earth is destroyed. We have maintained determinism because the outcome of the photon hitting the polarizer is predictable – the universe doubles. This idea is known as the "Many Worlds Interpretation of quantum mechanics." It was first proposed[5] in 1956 by Hugh Everett[6] (1930–1982).

The Schrödinger equation that we will meet in due course is utterly central to quantum mechanics. The time evolution of the Schrödinger equation is the time evolution of a physical system. One of the perplexing aspects of quantum mechanics is that the mathematical nature of the Schrödinger equation is deterministic, and so the time evolution of a physical system is deterministic. Within the Copenhagen interpretation of quantum mechanics, the deterministic Schrödinger equation suddenly becomes non-deterministic when we observe the physical system. Within the many worlds interpretation of quantum mechanics, the time evolution of the Schrödinger equation and thus of the universe remains deterministic.

The many worlds interpretation of quantum mechanics is not an idle speculation but is taken seriously by modern theoretical physicists. Of course, throughout the observable universe, there are trillions of photon/polarizer type events happening every microsecond, and so, according to the many worlds interpretation of quantum mechanics, there are trillions of new universes being created every microsecond.

---

[5] Everett, Hugh: Theory of the Universal Wavefunction. Thesis; Princeton University 1956 pp 1–140.
[6] Everett, Hugh: Relative State Formulation of Quantum Mechanics: Reviews of modern physics 29: 454–462.

5.2   A THOUGHT

Thinking back to a previous chapter, does determinism make any sense to an observer who is outside of time? The Euclidean complex plane, $\mathbb{C}$, has no time within it. If physics is written using these complex numbers, as wave physics must be, will that physics be deterministic?

CHAPTER **6**

THE NEWTONIAN WAVE EQUATION

In this chapter, we are concerned with only Newtonian mechanics. We will delve slightly more deeply into Newtonian mechanics that is usually done and we will discover special numbers (eigenvalues) within Newtonian mechanics. Newtonian mechanics does not overtly use complex numbers, but we will see that, when it comes to waves, Newtonian mechanics is led into the complex numbers.

Imagine a vibrating taut string stretched between two points that are a distance L apart – like a string on a guitar.

A Stretched String

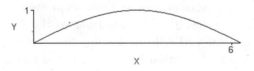

We assume that the string is ideal in that its gravitational mass and other extraneous properties can be ignored in the mathematics that follows. We also assume that $\frac{\partial y}{\partial x}$ is small. With those assumptions, the Newtonian motion of the ideal string is given by the Newtonian ideal wave equation:

$$\frac{\partial^2 y}{\partial x^2} - \frac{1}{v^2}\frac{\partial^2 y}{\partial t^2} = 0 \tag{6.1}$$

v is a constant incorporating the density and tension of the string.

This wave equation is subject to the boundary conditions that the vertical displacement, y, at the two ends, $x = \{0, L\}$ is zero. In mathematics, these boundary conditions are:

$$y(0, t) = y(L, t) = 0 \tag{6.2}$$

By applying forcing vibrations to the string, we discover that there are standing waves at particular resonances of the string such that the resonant frequencies correspond to half integral multiples of the string length.

Standing Waves

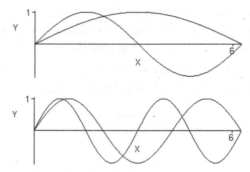

Since the string is described by the Newtonian wave equation, these standing waves are described by solutions of the Newtonian wave equation. Using standard differential equation techniques, we solve the Newtonian wave equation by assuming that there are solutions of the form:

$$y(x, t) = X(x)T(t) \tag{6.3}$$

Wherein X is a function of only x and T is a function of only t. We call these separable solutions. It is a fact that separable solutions of wave equations are standing wave solutions. Since standing waves are unchanging over time, we also refer to separable solutions as stationary solutions. Standing waves, stationary solutions, separable solutions – all the same thing.

Aside: We will use exactly this same method of finding standing wave solutions by assuming separable solutions when we solve the Schrödinger wave equation with a potential that is independent of time, but we will call the solutions stationary solutions rather than standing wave solutions. That is because, even though they are standing wave solutions, they do not change with time, and so they are, in that sense, stationary.

Putting the proposed solution into the Newtonian ideal wave equation leads to:

$$\frac{\partial^2 X(x)}{\partial x^2}\frac{1}{X(x)} = \frac{1}{v^2}\frac{1}{T(t)}\frac{\partial^2 T(t)}{\partial t^2} \tag{6.4}$$

The two variables are independent, but must be true for all values of $\{t, x\}$, and so both sides must be equal to the same constant, which, in anticipation of getting wave solutions, we call $-k^2$. (Without the minus sign, we get $\{\cosh(\ ), \sinh(\ )\}$ type solutions, which are nothing like waves. The squared bit avoids a square root sign in the solutions.)

$$\frac{\partial^2 X(x)}{\partial x^2}\frac{1}{X(x)} = -k^2$$

$$\frac{1}{v^2}\frac{1}{T(t)}\frac{\partial^2 T(t)}{\partial t^2} = -k^2 \tag{6.5}$$

With solutions:

$$X(x) = A\sin(kx) + B\cos(kx)$$
$$T(t) = C\sin(vkt) + D\cos(vkt) \tag{6.6}$$

Although we are doing purely Newtonian mechanics in this chapter, we have been driven to the $\{\sin(\ ), \cos(\ )\}$ trigonometric functions which properly exist in only the complex plane.

The boundary condition $X(0) = 0$ means that $B = 0$, and we have the solutions:

$$y(x,t) = \sin(k_n x)\left[C_n \sin(\omega_n t) + D_n \cos(\omega_n t)\right]$$

$$k_n = \frac{n\pi}{L}, \quad \omega_n = vk_n \tag{6.7}$$

The first four of these solutions, $n = 1, 2, 3, 4$, are the standing waves (stationary solutions or separable solutions) illustrated above.

The separable solutions of the Newtonian wave equation are the set of standing waves that are resonances of the ideal string. The reader should note that the standing waves exist only because the string is a fixed length. The equivalent to this in mathematical terms is that the standing wave solutions exist only because of the boundary conditions.

6.1   LINEAR DIFFERENTIAL EQUATIONS AND LINEAR SUMS AND LINEAR SPACES

A linear sum of mathematical objects, ϕ_i, is of the form:

$$\text{Linear Sum} = c_1\phi_1 + c_2\phi_2 + c_3\phi_3 + ...c_i\phi_i + ... \qquad (6.8)$$

Wherein $c_i \in \mathbb{C}$.

If a differential equation is such that it contains no terms which are higher than first powers of differentials, we say it is a linear differential equation. For example, if $\left(\dfrac{\partial y}{\partial x}\right)^2$ appears in the equation, then the equation is not linear, but we can have terms like $\dfrac{\partial^7 y}{\partial x^7}$ in a linear differential equation. The Newtonian ideal string wave equation is a linear differential equation, (and so is the Schrödinger equation). This definition is not important to us, and we give it only for completeness. The important thing about linear differential equations is that the solutions of linear differential equations are a linear space. In particular, the separable solutions of linear differential equations form a basis of the linear space of the solutions. We will deal with linear spaces in more detail later, but the centrally important properties of linear spaces are:

1.   Any linear sum of the elements (vectors or solutions to linear differential equations) that form the linear space is an element of that linear space.

And

2. Any element of a linear space is a linear sum of elements of that linear space.

3. Linear spaces have basis elements (basis solutions or basis vectors) which are orthogonal to each other. Linear spaces are often called vector spaces. In general, any complete set of vectors form a linear space.

The important bit is that the separable solutions of a linear differential equation are mutually orthogonal basis solutions. Any other solution can be written as a linear sum of only these solutions and any linear sum of the separable solutions of a linear differential equation is also a solution of that linear differential equation. It is exactly analogous to vectors in which we have a set of basis vectors that are orthogonal to each other and that span the whole vector space.

Because the Newtonian wave equation is linear, all the possible solutions of this equation, that is all possible shapes of the vibrating string (including the triangular profile of a plucked string), are a linear sum of basis solutions (also called separable solutions or standing waves or stationary solutions). It is a fact that the standing waves are a complete set of basis solutions of the Newtonian wave equation. Any other solution is a linear sum of these solutions. There are infinitely many standing waves, and so the other solutions might be an infinite sum of the standing wave solutions. We have the general solution of the Newtonian wave equation is:

$$y = \sum_n \sin\left(\frac{n\pi}{L}x\right)\left[C_n \sin\left(\frac{n\pi}{L}vt\right) + D_n \cos\left(\frac{n\pi}{L}vt\right)\right] \qquad (6.9)$$

This is a superposition (linear sum) of the standing waves.

Aside: The exact same situation applies to the Schrödinger wave equation. The separable solutions of the Schrödinger wave equation are the basis solutions of a linear space that contains all solutions as a linear sum of the basis solutions. We call the basis solutions of the Schrödinger wave equation stationary solutions because they correspond to standing waves – most often electrons in unchanging orbit around an atomic nucleus.

We see a concrete example of a linear space (vector space) in the Newtonian wave equation. The basis functions are the solutions that are standing waves. These solutions are orthogonal (under the overlap integral definition – see later) and complete – see later. We see that the basis solutions are a, infinite in number, discreet set. We see that the discreet set arose because of the imposition of the boundary conditions – the fixed length of the string.

If a linear differential equation has two solutions of the form:

$$y = \cos x$$
$$y = \sin x \tag{6.10}$$

Then, since any linear sum of solutions is also a solution, another solution is:

$$y = \cos x + i \sin x \tag{6.11}$$

Notice the imaginary coefficient of sin x. The linear sums of solutions of linear differential equations include sums with complex coefficients.

The reader is advised to re-read the last few paragraphs about linear spaces and basis solutions; when you understand this, you are well on the way to understanding quantum mechanics. All that we have done with the Newtonian wave equation, we will do with the Schrödinger equation.

6.2   FOURIER ANALYSIS

In the case of wave equations, we can calculate coefficients of the, usually infinite, linear sum of basis solutions of any desired solution using Fourier analysis. For example, if we want to know the solution of the Newtonian ideal string equation that corresponds to the initially triangular shape of a plucked string, or any other shape into which the string can be stretched, we can calculate that solution as a, most likely infinite, sum of the basis solutions using Fourier analysis. Fourier analysis is explained in many texts, and so we do not go into the details here.

The reader will have observed that when waves are superimposed upon each other, they form another, more complicated, wave. So it is with solutions of wave equations. When wave equation solutions are superimposed upon each other, they form another, more complicated, wave equation solution. A Fourier sum is just a linear superimposition (superposition) of waves.

6.3 NEWTONIAN EIGENVALUES AND NEWTONIAN COMPLEX NUMBERS

In the above Newtonian wave equation, we see there are special numbers (eigenvalues) associated with the standing wave solutions. These special numbers are $k_n = \dfrac{n\pi}{L}$.

6.4 TWO WAYS OF SEEING THE NEWTONIAN WAVE EQUATION

The stationary solution to the Newtonian wave equation, (6.7), is:

$$y(x,t) = \sin(k_n x)\left[C_n \sin(\omega_n t) + D_n \cos(\omega_n t)\right] \qquad (6.12)$$

The time independent part of this is:

$$y(x) = \sin(k_n x) \qquad (6.13)$$

We have:

$$\frac{\partial^2 y}{\partial x^2} = -k_n{}^2 \sin(k_n x) \qquad (6.14)$$

We can see the double differentiation as a mathematical operator that acts on a function to turn it into another function. In the case above, the double differentiation operator has acted upon the $\sin(k_n x)$ function to turn it into a multiple of itself, $-k_n^2$. Mathematicians say that the $\sin(k_n x)$ function is an eigenfunction (special function) of the double differentiation operator and that associated with that eigenfunction is the eigenvalue (special number) $-k_n^2$.

If, instead of k_n, we have a particular number, say $\dfrac{3}{L}$, then $\sin\left(\dfrac{3x}{L}\right)$ is an eigenfunction of the double differentiation operator with eigenvalue $-\dfrac{9}{L^2}$.

Alternatively, the reader can look at the equation (6.14) in the way we normally look at it. Quantum mechanics chooses the operator, eigenfunction, and eigenvalue way of looking at such an equation.

6.5   NON-STATIONARY STATES

The standing wave solution (6.7) is a solution for any values of $\{C_n, D_n\}$. Let us consider $C_n = i$ & $D_n = 1$. We then have:

$$y(x,t) = \sin(k_n x)\big[i\sin(\omega_n t) + \cos(\omega_n t)\big] \qquad (6.15)$$

The modulus of this is:

$$\big|y(x,t)\big|^2 = \sin^2(k_n x) \qquad (6.16)$$

We see that the modulus is independent of time. Now let us consider a sum of two different standing wave solutions. We have

$$
\begin{aligned}
y(x,t) &= \sin(k_m x)\big[i\sin(\omega_m t) + \cos(\omega_m t)\big] \\
&\quad + \sin(k_n x)\big[i\sin(\omega_m t) + \cos(\omega_m t)\big] \\
&= \cos(\omega_m t)\sin(k_m x) + \cos(\omega_m t)\sin(k_n x) \\
&\quad + i\big(\sin(\omega_m t)\sin(k_m x) + \sin(\omega_m t)\sin(k_n x)\big)
\end{aligned}
\qquad (6.17)
$$

The modulus of this is:

$$
\begin{aligned}
\big|y(x,t)\big|^2 &= \sin^2(k_m x) + \sin^2(k_n x) \\
&\quad + \sin(k_m x)\sin(k_n x)\big[\cos((\omega_n - \omega_m)t)\big]
\end{aligned}
\qquad (6.18)
$$

Which is not independent of time.

We will meet something very similar to this in quantum mechanics. In general, within quantum mechanics, the observable quantity (probability density) is the modulus squared of a complex number, $|\psi|^2$ In general, within quantum mechanics, a standing wave solution corresponds to a physical state that is unchanging in time and a sum of standing wave solutions corresponds to a physical state that is changing in time. A moving particle is a physical system that is changing in time. In general, within quantum mechanics, particles that are bound within a potential well, such as an electron orbiting an atom, do not change over time and are described by standing wave solutions of the Schrödinger wave equation whereas particles that are not bound within a potential well (freely moving particles) are described by sums of standing wave solutions of the Schrödinger wave equation. Within quantum mechanics, we could say that the property that defines a basis solution (standing wave solution), ψ_i, is that its probability density, $|\psi_i|^2$, is independent of time. The sums of basis solutions are such that their probability densities are not independent of time.

$$
\begin{aligned}
\text{Unchanging physical systems} &= \text{A single basis solution} \\
\text{Changing physical systems} &= \text{A sum of basis solutions}
\end{aligned}
\tag{6.19}
$$

A moving particle is a changing physical system. An electron in (undisturbed) orbit is an unchanging physical system.

6.6   SUMMARY

The basis solutions of the Newtonian wave equation are found to be the separable solutions also called stationary solutions or standing waves. All other solutions are linear sums of these solutions. All linear sums of these solutions are also solutions. Boundary conditions "cause" these solutions to exist. Each basis solution is associated with a special number (eigenvalue).

6.7   A LOOK AHEAD

We will in due course solve the Schrödinger wave equation by separation of variables. This can always be done provided we have a potential that is independent of time. Boundary conditions will lead to basis solutions associated with eigenvalues. The eigenvalues will correspond to standing waves. The standing waves are (usually) electron orbits in which in which the circumference of the orbit is a multiple of the wavelength of the electron, $n\lambda = 2\pi r$. Electrons do not orbit a nucleus like planets orbit the sun; electrons are just standing waves that fit around the nucleus in such a way that the ends of the waves fit together to form a standing wave. Since an electron has a particular wavelength, electrons can occupy only particular discreet orbits that correspond to an integral number of wavelengths. Thus, electrons in orbit around an atomic nucleus will have only allowed discreet amounts of orbital angular momentum and only allowed discreet amounts of energy. In fact, electron orbital angular momentum is multiples of Planck's constant divided by 2π, $L = n\hbar$.[1]

Aside: It is sometimes confusing that the basis solutions of a wavefunction are referred to by many different names within physics. The various names include:

Bound State Wavefunctions	Bound States
Stationary States	Eigenstates
Basis Eigenfunctions	Standing Waves
Basis Solutions	Basis Vectors

6.8   AUTHOR'S NOTE

The next two chapters are the most difficult chapters in this book. The reader will need to read and re-read these two chapters. Once the reader has mastered the next two chapters, the remainder of this book will be an easy read. Well, not quite that easy, but a lot easier than the next two chapters.

[1.] This is where the n's in quantum mechanics comes from. We say, "Planck put the n is quantum."

LINEAR SPACES AND INNER PRODUCTS

Aside: The axioms of a linear space, V, are:

1. V is closed under addition. This means that the sum of any two elements of the linear space is also an element of the linear space.

2. Multiples (by scalars) of elements of V are elements of V. In quantum mechanics, the scalars are complex numbers rather than real numbers – they could have been quaternions perhaps.

3. Addition within V is commutative.

4. Addition is associative.

The reader does not need to know the details of what these axioms mean; indeed, other authors might, entirely justifiably, list seven axioms of a linear space rather than just four – it is a matter of opinion as to how many we need. We include this list of axioms for presentational completeness only.

It is a quite obvious fact that the complex numbers, $\mathbb{C}$, are the linear sum of two basis elements. By this, we mean that any complex number can be written as the sum of multiples of these basis elements:

$$\begin{bmatrix} a & b \\ -b & a \end{bmatrix} = a\begin{bmatrix} 1 & 0 \\ 0 & 1 \end{bmatrix} + b\begin{bmatrix} 0 & 1 \\ -1 & 0 \end{bmatrix}$$
$$\{a,b\} \in \mathbb{R}$$

(7.1)

The numbers $\{a, b\}$ are real numbers, and technically, they ought to be written as 2×2 homothetic matrices, but the idea is clear. All division algebras can be written as the linear sum of a finite number of basis elements. The hyperbolic complex numbers are:

$$\begin{bmatrix} h & 0 \\ 0 & h \end{bmatrix}\begin{bmatrix} \cosh\chi & \sinh\chi \\ \sinh\chi & \cosh\chi \end{bmatrix} =$$
$$\begin{bmatrix} h & 0 \\ 0 & h \end{bmatrix}\begin{bmatrix} \cosh\chi & 0 \\ 0 & \cosh\chi \end{bmatrix} + \begin{bmatrix} h & 0 \\ 0 & h \end{bmatrix}\begin{bmatrix} 0 & \sinh\chi \\ \sinh\chi & 0 \end{bmatrix}$$

(7.2)

$$\{h,\chi\} \in \mathbb{R}$$

In this case, we have been technically correct and written the real number coefficients as 2×2 homothetic matrices. The quaternions are, in non-matrix notation:

$$a + \hat{i}b + \hat{j}c + \hat{k}d = a + b(\hat{i}) + c(\hat{j}) + d(\hat{k})$$
$$\{a,b,c,d\} \in \mathbb{R}$$

(7.3)

7.1   COMPLETENESS

Not only can every complex number be written as the linear sum of two basis elements, but every linear sum of these two basis elements is a complex number. We say that the complex numbers are "complete" because every complex number can be written as a linear sum of the two basis elements and every linear sum of the two basis elements is a complex number. We need both conditions to be satisfied to have completeness.

Every division algebra has this property of completeness. There are things other than division algebras that have this property of completeness. One example is the set of solutions of a linear differential equation. Of all the solutions of a linear differential equation, some are basis solutions.

Aside: We saw above that the standing wave solutions of the New-tonian ideal wave equation are such basis solutions. The Newtonian wave equation for an ideal vibrating string is the linear differential equation:

$$\frac{\partial^2 y}{\partial x^2} - \frac{1}{v^2}\frac{\partial^2 y}{\partial t^2} = 0$$

(7.4)

For a vibrating string with both ends fixed, this equation has solutions which are standing waves with wavelengths corresponding to $\left\{\frac{L}{2}, L, \frac{3L}{2}, ...\right\}$ where L is the length of the string. These standing wave solutions are orthogonal to each other and are complete in that all the possible solutions of the equation for a fixed string of length L can be written as a linear sum of the standing wave solutions and every linear sum of the standing wave solutions is a solution of the equation. The standing wave solutions of the Newtonian ideal fixed length string differential equation are basis solutions.

We can form a linear sum of the basis solutions of a linear dif-ferential equation and that sum will be a solution of the linear differ-ential equation. Furthermore, any solution of the linear differential equation will be a linear sum of the basis solutions of that linear dif-ferential equation. If the basis solutions are φ_i, then all solutions of the linear differential equation are of the form:

$$\Phi = c_1\varphi_1 + c_2\varphi_2 + c_3\varphi_3 + ...$$
$$c_i \in \mathbb{C}$$

(7.5)

Above, within the division algebra, $\mathbb{C}$, we have taken the coefficients of the basis elements to be real numbers. For linear differential equations, the coefficients need not be real numbers, they can be complex numbers or, indeed, they can be elements of any division algebra ($\{\mathbb{R}, \mathbb{C}, \mathbb{H}, \mathbb{S}, A_3...\}$. In division algebras, we always take the coefficients to be real numbers.

7.2 DIMENSION

The real numbers, $\mathbb{R}$, are a 1-dimensional division algebra. The complex numbers, $\mathbb{C}$, are a 2-dimensional division algebra. The quaternions, $\mathbb{H}$, are a 4-dimensional division algebra. The dimension of a division algebra is the number of basis elements. This concept of dimension can be applied to the solutions of a linear differential equation. If there are n basis solutions, the set of solutions is said to be n-dimensional. A division algebra is a geometric space. The solutions of a linear differential equation are not a geometric space – they are a set of solutions – but the concept of dimension can be associated with that set.

Aside: We are transplanting the spatial concept of dimension into a non-spatial set of solutions. Is this allowed? It seems to work in quantum mechanics, and so we brush philosophical considerations under the carpet and plough onward.

Within a division algebra, we can multiply two elements of that algebra together to form another element of that algebra. We say the division algebra is multiplicatively closed[1]. If we multiply two solutions of a linear differential equation together, in general, we do not get another solution of that linear differential equation. The solutions of a linear differential equation are not multiplicatively closed. The set of solutions does not have the property of multiplicative closure, but this simply means that the solutions of a linear differential equation are not a division algebra. They are only a linear space.

7.3   ORTHOGONALITY

No matter how hard I try, I cannot write one of the basis elements of the complex numbers as a linear sum of the other basis element:

$$a\begin{bmatrix} 1 & 0 \\ 0 & 1 \end{bmatrix} \neq b\begin{bmatrix} 0 & 1 \\ -1 & 0 \end{bmatrix} \quad \forall a \in \mathbb{R} \qquad (7.6)$$

We say that the basis elements are orthogonal. We mean that each of them cannot be written as a linear sum of the other basis elements.

---

[1] Any multiplicatively closed set of non-singular matrices is a division algebra. The dimension of the algebra is the size of the matrices. The non-singularity is most often gained by exponentiating the multiplicatively closed matrices.

The reader will often see orthogonality defined in terms of an inner product. It can be done this way, but we do not need to do it that way; we do not need an inner product to define orthogonality. We can define orthogonality with no more than the concept of a linear sum of basis elements.

The set of solutions of a linear differential equation includes basis elements (basis solutions) that are orthogonal to each other in that each of them cannot be written as a linear sum of the other basis solutions.

7.4 VECTORS

There are other things besides the set of solutions of a linear differential equation and the division algebras that have these properties of completeness, dimension, and orthogonal basis elements. The set of ordered n-tuples of real numbers (we call these vectors) are such that every one of them can be written as a linear sum of n basis elements and that any linear sum of these basis elements is a member of the set. We demonstrate with $\mathbb{R}^3$:

$$\begin{bmatrix} a \\ b \\ c \end{bmatrix} = a \begin{bmatrix} 1 \\ 0 \\ 0 \end{bmatrix} + b \begin{bmatrix} 0 \\ 1 \\ 0 \end{bmatrix} + c \begin{bmatrix} 0 \\ 0 \\ 1 \end{bmatrix} \quad : \quad \{a,b,c\} \in \mathbb{R} \qquad (7.7)$$

Another example, and one that is central to quantum mechanics, is the set of ordered n-tuples of complex numbers (we call these complex vectors). We demonstrate with $\mathbb{C}^3$:

$$\begin{bmatrix} a+id \\ b+ie \\ c+if \end{bmatrix} = a \begin{bmatrix} 1 \\ 0 \\ 0 \end{bmatrix} + d \begin{bmatrix} i \\ 0 \\ 0 \end{bmatrix} + b \begin{bmatrix} 0 \\ 1 \\ 0 \end{bmatrix} + e \begin{bmatrix} 0 \\ i \\ 0 \end{bmatrix} + c \begin{bmatrix} 0 \\ 0 \\ 1 \end{bmatrix} + f \begin{bmatrix} 0 \\ 0 \\ i \end{bmatrix} \qquad (7.8)$$

$$\{a,b,c,d,e,f\} \in \mathbb{R}$$

In all these cases, we see that any vector, real or complex, can be written as a linear sum of the appropriate basis elements and that any linear sum of these basis elements is a vector, real or complex respectively, and that none of the basis elements can be written as a linear sum of the other basis elements - we see completeness and orthogonality.

Any set of mathematical objects that is complete and has orthogonal basis elements is called a linear space. Another name for a linear space is a vector space. It is all down to linear sums of basis elements.

7.5   LINEAR SPACES (VECTOR SPACES) AND QUANTUM MECHANICS

In quantum mechanics, the state of a physical system (perhaps a particle) is represented by an element in a linear space. (There is usually more than one basis to the linear space.) The elements of a linear space are often called vectors (sometimes, they are vectors in a space of infinite dimension). Thus, in quantum mechanics, the state of a physical system is a vector.

Since we take an unobserved physical system to be in a superposition (linear sum) of all possible states, then associated with that superposition are the coefficients of the basis states in the linear sum. These coefficients are an n-tuple of complex numbers – a complex vector. If, as is the case with non-commuting operators, there is more than one basis of the linear space involved, the state is one vector written in more than one basis. When the state is observed, it collapses into a basis state (a basis vector in a particular basis), but this is still a vector; it is, however, only one number (the other numbers in the n-tuple are zero). In quantum mechanics, we write vectors as:

$$\text{vector} = |\psi\rangle \tag{7.9}$$

These vectors are also called kets.

We will find that each quantum mechanical operator is associated with a set of basis eigenfunctions that are the solutions of an eigenvalue equation[2] containing the operator and the basis eigenfunctions are the basis elements of a linear space. This is another way of saying that the set of eigenfunctions (eigensolutions) which describe the possible states of a physical system (think possible states of an electron orbiting an atomic nucleus) is a linear space. The elements of the lin-

_____
[2.] We will cover eigenvalue equations shortly.

ear space are superpositions (linear sums) of the basis eigenfunctions – just like vectors are linear sums of basis vectors.

The dimension of the set, that is the number of basis elements, can be infinite. This rather strains the brain, and the details are beyond this book.

7.6   COLLAPSE INTO A BASIS SOLUTION

The different energy levels of an electron orbiting an atomic nucleus are each associated with a different basis solution (basis vector or standing wave) of an appropriate linear differential equation called the Schrödinger equation for that system. Prior to observation, the electron is seen as being in a state described by a superposition (linear sum) of all the basis solutions of the appropriate Schrödinger equation. When the electron is observed, it "collapses" into a state described by a single basis solution. Prior to observation, the electron is a sum of standing waves; when observed, it becomes a single standing wave; after some time has passed, it has evolved from being a single standing wave into again being a superposition of standing waves.

If there were, say, three possible orbits that an electron might occupy, $\{\varphi_1, \varphi_2, \varphi_3\}$, then, prior to observation it would be described by the wavefunction (sum of standing waves):

$$\Psi = c_1\varphi_1 + c_2\varphi_2 + c_3\varphi_3$$
$$c_i \in \mathbb{R} \tag{7.10}$$

When observed, it would "collapse" into only one of these basis solutions (basis states or standing waves), say φ_3, and would then be described by the wavefunction (standing wave):

$$\Psi = \varphi_3 \tag{7.11}$$

The "choice" of which of the three basis solutions (also called eigenstates) the wavefunction collapses into is purely a matter of chance. The probability of it being the eigenstate φ_i is determined by the relative size of the modulus of the complex coefficient of that eigenstate,. $|c_i|^2 = c_i^* c_i$, compared to the moduli of the other coefficients.

Standing wave, basis state, eigenstate, basis vector, basis solution, eigensolution – all the same thing!

7.7 INNER PRODUCTS

Inner products are a central part of quantum mechanics. The idea behind them derives from the inner products of division algebras, but a quantum mechanical inner product is both a very different thing and a very similar thing from and to a division algebra inner product.

The nature of an inner product within a division algebra is not part of quantum mechanics[3]; none-the-less, we include a description of such division algebra inner products as an aside for the edification of the reader. We choose to explain the inner product in the division algebra $\mathbb{C}$.

Aside: Because a division algebra is a geometric space, it has angles, trigonometric functions, and a rotation matrix. A number within a division algebra is a position in the geometric space of that algebra – think complex plane. A position in a geometric space is a displacement vector, and so numbers and vectors are the same thing within a division algebra.

Within a division algebra, the inner product is a way of calculating the angle between two vectors subtended at the origin. In general, within a division algebra, we calculate the angle between two vectors by conjugating one of the vectors and taking the product of the conjugated vector with the other vector. We demonstrate with unit length complex numbers, $\mathbb{C}$, in polar form (look for the position of the minus sign).

$$\begin{bmatrix} \cos\theta & -\sin\theta \\ \sin\theta & \cos\theta \end{bmatrix}\begin{bmatrix} \cos\phi & \sin\phi \\ -\sin\phi & \cos\phi \end{bmatrix}$$
$$= \begin{bmatrix} \cos(\theta-\phi) & \sin(\theta-\phi) \\ -\sin(\theta-\phi) & \cos(\theta-\phi) \end{bmatrix}$$

(7.12)

[3.] They are of central importance within special relativity.

We see that by multiplying a conjugated vector with another vector, we get a measure of the angle $(\theta - \phi)$ between the two vectors in the form of a trigonometric function with the angle between the vectors as its argument. In Cartesian form, this is:

$$\frac{1}{\sqrt{s^2 + t^2}} \begin{bmatrix} s & t \\ -t & s \end{bmatrix} \bullet \frac{1}{\sqrt{x^2 + y^2}} \begin{bmatrix} x & y \\ -y & x \end{bmatrix}$$

$$= \frac{1}{\sqrt{s^2 + t^2} \sqrt{x^2 + y^2}} \begin{bmatrix} s & -t \\ t & s \end{bmatrix} \begin{bmatrix} x & y \\ -y & x \end{bmatrix} \qquad (7.13)$$

$$= \frac{1}{\sqrt{s^2 + t^2} \sqrt{x^2 + y^2}} \begin{bmatrix} sx + ty & sy - tx \\ -(sy - tx) & sx + ty \end{bmatrix}$$

Putting these equal, we get:

$$\cos(\theta - \phi) = \frac{sx + ty}{\sqrt{s^2 + t^2} \sqrt{x^2 + y^2}} \qquad (7.14)$$

Which the reader will doubtless recognize as being the same as the conventional dot-product of two 2-dimensional $\mathbb{R}^2$ vectors.

We also get the magnitude of the vector cross-product[4] of two vectors as a bonus.

$$\sin(\theta - \phi) = \frac{sy - tx}{\sqrt{s^2 + t^2} \sqrt{x^2 + y^2}} \qquad (7.15)$$

In case the reader was wondering, the cross product is called the outer product.

In space-time (the hyperbolic complex numbers), the angle between two space-time vectors is the argument of the cosh( ) function.

In the above aside, we have introduced the concept of the inner product of one complex number with another as the conjugate of one number multiplied by the other number. Within a division algebra, the inner product makes perfect sense as a measure of the angle between two numbers, but the solutions of a linear differential equation are not a geometric space and do not have angles between

---

[4.] Note that this is not a vector sticking out of the complex plane, $\mathbb{C}$. The complex numbers are a 2-dimensional algebra; they cannot "grow" another dimension any more than the reader can grow another arm.

them. Nor are ordered n-tuples of real numbers a geometric space. We are shortly going to put an inner product on to ordered n-tuples of real numbers and on to ordered n-tuples of complex numbers and on to solutions of a linear differential equation. Since these sets of mathematical objects have no concept of angle within them, we are going to have to drop the association between an inner product and an angle.

If the inner product of two complex numbers is zero, the cosine of the angle between the two vectors is zero. A zero cosine corresponds to an angle of 90° within the complex plane, $\mathbb{C}$. So, if the inner product of two vectors is zero, they are perpendicular to each other. If two vectors in the complex plane are perpendicular to each other, the two vectors are such that we cannot write one as a linear sum of the other – the two vectors are orthogonal. We have a correlation between zero inner product and orthogonality[5]. We will carry this with us into quantum mechanics.

If the inner product of two normalized complex numbers is unity, the cosine of the angle between the two vectors is unity. A unity cosine corresponds to an angle of 0° within the complex plane, $\mathbb{C}$. So, if the inner product of two normalized vectors is unity, they are the same vector. We have a correlation between unity inner product and identity of the vectors. We will carry this with us into quantum mechanics.

7.8   INNER PRODUCTS IN $\mathbb{R}^n$

Within $\mathbb{R}^3$, we seek a way of combining two ordered 3-tuples of real numbers (two 3-dimensional vectors) in such a way that the combination is zero if we cannot write either one of the 3-tuples as a linear sum of the other and the combination is unity if the normalized 3-tuples are the same.

It is customary to call this way of combining two normalized 3-tuples of three real numbers the inner product on $\mathbb{R}^3$. We all know what it is:

---

[5] Within space-time, the cosh( ) is never zero, and so not all division algebras have a correlation between orthogonality and zero inner product.

$$\left\langle \begin{bmatrix} \dfrac{a}{\sqrt{a^2+b^2+c^2}} \\ \dfrac{b}{\sqrt{a^2+b^2+c^2}} \\ \dfrac{c}{\sqrt{a^2+b^2+c^2}} \end{bmatrix} , \begin{bmatrix} \dfrac{x}{\sqrt{x^2+y^2+z^2}} \\ \dfrac{y}{\sqrt{x^2+y^2+z^2}} \\ \dfrac{z}{\sqrt{x^2+y^2+z^2}} \end{bmatrix} \right\rangle \qquad (7.16)$$

$$= \dfrac{ax+by+cz}{\sqrt{a^2+b^2+c^2}\sqrt{x^2+y^2+z^2}}$$

In the above, we have used the $\langle,\rangle$ notation to indicate the inner product. This notation is ubiquitous within quantum mechanics. Notice that the denominators of the components of the vectors normalize the vector.

Not only does this way of combining two mathematical objects work in $\mathbb{R}^3$, but analogous ways of combining two normalized ordered n-tuples of real numbers works all the way up to infinity. We are therefore able to impose this way of combining two ordered n-tuples of real numbers on to the vector spaces of $\mathbb{R}^n$ and call it the inner product of those vector spaces.

7.9   INNER PRODUCTS IN $\mathbb{C}^n$

Quantum mechanics uses complex vectors in $\mathbb{C}^n$. These are ordered n-tuples of complex numbers. There is a way of combining two normalized complex vectors such that, when the combination is zero, the vectors are orthogonal (think sums of basis vectors), and that, when the combination is unity, the vectors are the same.

This way of combining two such complex vectors is called the inner product on $\mathbb{C}^n$. It is very similar to the inner product within the complex plane in that we conjugate every element of one vector and then multiply the respective pairs of elements together and sum the lot:

$$\left\langle \begin{bmatrix} a+ib \\ c+id \end{bmatrix} , \begin{bmatrix} e+if \\ g+ih \end{bmatrix} \right\rangle = \begin{bmatrix} a-ib \\ c-id \end{bmatrix} \bullet \begin{bmatrix} e+if \\ g+ih \end{bmatrix}$$

$$\qquad (7.17)$$

$$= (a-ib)(e+if)+(c-id)(g+ih)$$

$$= (ae+bf+cg+dh)+i(af-be+ch-dg)$$

Notice that, if we swap the vectors within the inner product, we get the conjugate of the above:

$$\left\langle \begin{bmatrix} e+if \\ g+ih \end{bmatrix}, \begin{bmatrix} a+ib \\ c+id \end{bmatrix} \right\rangle = (ae+bf+cg+dh) - i(af-be+ch-dg) \quad (7.18)$$

The reader will often see the $\mathbb{C}^n$ inner product expressed as:

$$\left\langle \begin{bmatrix} e+if \\ g+ih \end{bmatrix}, \begin{bmatrix} a+ib \\ c+id \end{bmatrix} \right\rangle = [e-if \quad g-ih] \begin{bmatrix} a+ib \\ c+id \end{bmatrix} \quad (7.19)$$

Wherein the inner product appears as matrix multiplication. For higher dimensional n-tuples of complex numbers, the inner product is of the same form.

The conjugate transpose is often called a dual vector or a bra vector and is written as:

$$[e-if \quad g-ih] = \langle \psi | \quad (7.20)$$

This is combined with the ket vector and written as a bracket[6]:

$$\langle \psi | \phi \rangle \equiv [e-if \quad g-ih] \begin{bmatrix} a+ib \\ c+id \end{bmatrix} \quad (7.21)$$

We sometimes see the inner product of two complex vectors, $\{A, B\} \in \mathbb{C}_n$ written as:

$$\langle A, B \rangle = A^\dagger B \quad (7.22)$$

7.10 INNER PRODUCTS IN QUANTUM MECHANICS

Quantum mechanics does not use the concept of angle between complex vectors, but it does use the concept of an inner product being unity or zero if the complex vectors are the same or orthogonal respectively. Quantum mechanics copies these concepts from the

---

[6.] Paul Dirac is the origin of both these names and the sense of humor that goes with them.

complex number division algebra while leaving behind the concept of angle and geometric space.

Within quantum mechanics, we use the inner product to calculate the probability that a system which is in a superposition, Ψ, of basis states, ϕ_i, will be observed to be in a particular one of those basis states. Repeating the above, the superposition is a linear sum, with complex coefficients, of the possible basis states:

$$\Psi = c_1\phi_1 + c_2\phi_2 + c_3\phi_3 + ...$$
$$c_i \in \mathbb{C}$$

(7.23)

We uniformly scale all the coefficients to the point where the sum of their moduli is unity. We call this normalization. The probability of observing the system to be in the state ϕ_i is the modulus of the normalized complex coefficient, c_i. Clearly, $|c_i| = c_i^{\circ} c_i \leq 1$. We seek a way of combining a particular basis state with the superposition of basis states to produce the modulus of the coefficient of the basis state within the superposition - we want a mathematical way to pick out the appropriate coefficient. This is very similar to using the inner product of a basis vector and a general vector in $\mathbb{R}^n$ to pick out a particular component of a vector.

$$\begin{bmatrix} 1 \\ 0 \\ 0 \end{bmatrix} \bullet \begin{bmatrix} a \\ b \\ c \end{bmatrix} = \begin{bmatrix} 1 & 0 & 0 \end{bmatrix} \begin{bmatrix} a \\ b \\ c \end{bmatrix} = [a]$$

(7.24)

The basis solutions of a linear differential equation are functions. We therefore need a way of picking out the coefficient of one of the basis functions from the superposition of basis functions. We want an inner product of functions.

7.11   INNER PRODUCTS OF FUNCTIONS

We can combine two normalized functions together in such a way that, when the combination is equal to unity, the normalized functions are the same function, and, when the combination is zero, the functions are orthogonal. In particular, we can do it with solutions of linear differential equations. In the case of functions, which can be complex, the inner product is an overlap integral.

$$\langle \psi, \phi \rangle = \int_{-\infty}^{\infty} dx \ \psi^{\circ} \phi \qquad (7.25)$$

This is analogous to a vector inner product in a space of infinite dimension, $\mathbb{C}^{\infty}$. Only if the functions form a vector space (linear space) can we apply this inner product.

A linear space with an inner product is called a Hilbert space after the mathematician David Hilbert.

Aside: David Hilbert (1862–1943) was one of the most prominent mathematicians of the early 20th century. In the path of Euclid, Hilbert advocated formalizing mathematics as a series of axioms, and he made his own contribution in the form of twenty (originally twenty-one) axioms that formalized geometry[7]. This idea is known as Hilbert's program. In 1902, he addressed the International Congress of Mathematics in Paris with a list of the central problems of mathematics which determined the direction of mathematics for much of the 20th century. To the physics student, the concept of a Hilbert space (linear space) is his greatest contribution.

Building upon the reputations of Gauss, Riemann, Dedekind, and Dirichlet, Hilbert made Göttingen University into a world renowned center for mathematics. Unfortunately, many of the mathematicians in Göttingen in the 1930's were Jewish, and the Nazis effectively destroyed the mathematics department there during the 1930's. Hilbert is well known for his opposition to the Nazis and for his support in the earlier part of the 20th century of the brilliant female mathematician Emmy Noether against the anti-women prejudices of the university officials at the time.

The geometries of the n-dimensional division algebras with one real axis and $n-1$ imaginary axes which are widely mentioned in this book are an entirely different view of geometry from the one advocated by Hilbert.

[7.] David Hilbert: The Foundations of Geometry (1902).

7.12 OVERLAP INTEGRALS

The overlap integral is taken to be the inner product of two functions within a linear space. The product of two functions is non-zero only when both functions are non-zero, Thus the product is a measure of the "overlap" of the two functions.

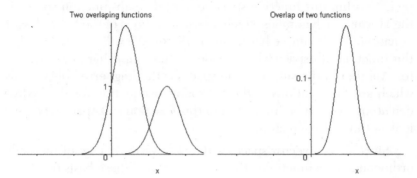

We form the overlap integral to sum the product of the functions at every point. To make this work, we need to normalize the area under the functions (the integral of the functions). The overlap integral is much used in quantum mechanics.

7.13 LINEAR SUB-SPACES

Linear spaces have sub-spaces. From the set of basis elements, we select, at random, a few. The linear sums of these few basis elements will form a linear space in their own right. It is like $\mathbb{R}^2$ is a sub-space of $\mathbb{R}^3$. So it is with the solutions of a linear differential equation; if we select a few basis solutions, the linear sums of these few basis solutions are a vector space in their own right. This is very different from the sub-spaces of the division algebras whose existence is very much restricted by the nature of the finite group that underlies the algebra.

7.14 SQUARE INTEGRABLE FUNCTIONS

There is a large set of functions that form a function space that is central to quantum mechanics because it contains all the possible functions that are within quantum mechanics. That large set of functions is called the set of square integrable functions and is denoted by L^2. Within this function space, one of the sub-function spaces is the Hermite polynomials which are the solutions of the Schrödinger equation for the simple harmonic oscillator. We will be interested in this function sub-space when we look at the simple harmonic oscillator. Another of the sub-function spaces is the Laguerre polynomials which are the solutions of the Schrödinger equation for the hydrogen atom. We will be interested in this function sub-space when we look at the hydrogen atom.

Since L^2 is a vector space, within L^2, we have a set of mutually orthogonal basis functions. Every linear sum of these basis functions is a square integrable function, and every square integrable function is a linear sum of these basis functions. The inner product of the square integrable functions is the overlap integral above, (7.25).

7.15 SQUARE INTEGRABILITY

The integral of a function is the area between the graph of the function and the axis. Area below the axis counts as negative, and area above the axis counts as positive. It is possible to have both positive area (above the axis) and negative area (below the axis) in such amounts that the total area is zero even though there is area between the graph of the function and the axis. The cosine function between $[0, 2\pi]$ is such a function. If we square the function before integrating, then all the area between the graph of the squared function and the axis will be positive, and this is a better measure of the integral of the function because it avoids the cancellation of positive area by other negative area.

A function is said to be square integrable if the area between the graph of its square and the axis is finite; if this area is infinite, the function is not square integrable.

A Function that is Square Integrable

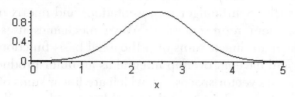

A Function that is not Square Integrable

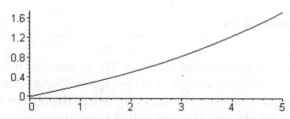

The set of square integrable functions is denoted by:

$$L^2(\mathbb{R}^3): \left\{ \psi : \mathbb{R}^3 \to \mathbb{C} : \int_{\mathbb{R}^3} dx \, |\psi(x)|^2 < \infty \right\} . \qquad (7.26)$$

This set is a linear space with an inner product (a Hilbert space).

Square integrable functions are:

i. Defined everywhere.

ii. Infinitely differentiable. We can differentiate them time and time again until they are just zero.

iii. Continuous everywhere.

iv. Have continuous derivatives.

v. Have finite square integrals. They do not zoom off to infinity thereby having an infinite area beneath their graphs.

7.16 THE TWO REPRESENTATIONS OF QUANTUM MECHANICS

Quantum mechanics is formulated in two ways known as wave mechanics (the Schrödinger representation) and matrix mechanics (the Heisenberg representation). Wave mechanics uses function spaces which are linear sums of orthogonal basis functions and are complete and have an overlap integral as an inner product. Matrix mechanics uses vector spaces, $\mathbb{C}^n$, which are linear sums of orthogonal basis vectors and are complete and have a "dot-product" as an inner product.

7.17 SUMMARY

This has not been an easy chapter. We summarize it briefly and advise the reader to re-read it. A vector space (linear space) is a set of mathematical objects such that every one of the objects is a linear sum of "special objects" that are called basis objects and such that any linear sum of these basis objects is a member of the set; this is called completeness. The "special objects" are such that none of them can be written as a linear sum of the others of them; this is called orthogonality.

A vector space with an inner product[8] is called a Hilbert space.

It is quite remarkable that sets of functions and sets of ordered n-tuples of real numbers and sets of ordered n-tuples of complex numbers should have these properties of basis functions (basis vectors), orthogonality, and completeness. It is quite remarkable that there exist inner products of functions and of vectors that are utterly divorced from the geometric spaces of the division algebras that inspired them. Even so, these concepts are central to quantum mechanics.

[8.] Technically, an inner product has to satisfy various axioms; it cannot be any old thing.

7.18 BRAS AND KETS (ADDENDUM)

We assume the existence of an orthonormal basis of vectors such that the inner product of these basis vectors is either unity or zero. We write this as:

$$\left\langle \phi_i \middle| \phi_j \right\rangle = \delta_{ij}$$

$$|\psi\rangle = c_1 |\phi_1\rangle + c_2 |\phi_2\rangle + c_3 |\phi_3\rangle + \dots \tag{7.27}$$

To convert a ket vector, $|\phi\rangle$, into a bra vector, $\langle\phi|$, we complex conjugate the components of the ket vector:

$$c|\psi\rangle = |c\psi\rangle \rightarrow c^\circ \langle\psi| \tag{7.28}$$

EXERCISES

1. Is the function $f(x) = e^{ax}$ square integrable?

2. Is the function $f(x) = e^{-ax^2}$ square integrable?

3. If two vectors in $\mathbb{C}^3$ are:

$$A = \begin{bmatrix} 2 \\ 3+i \\ 7i \end{bmatrix} \quad \& \quad B = \begin{bmatrix} i \\ 2 \\ 4 \end{bmatrix} \tag{7.29}$$

   What are the inner products $\langle A, B \rangle$ & $\langle B, A \rangle$?

4. Is $f(x) = \cosh(x)$ square integrable?

5. Is $f(x) = e^{-x}\cosh(x)$ square integrable?

6. Is $f(x) = e^{-x^2}\cosh(x)$ square integrable?

CHAPTER 8

OPERATORS, EIGENFUNCTIONS, AND EIGENVALUES

Is the reader still sitting comfortably? Then we shall continue. Mathematically, quantum mechanics is formulated in terms of linear operators, eigenfunctions, and eigenvalues. Technically, a linear operator changes (maps if you prefer) an element (vector or function) of a linear space into another element of that linear space. In quantum mechanics, the basis functions of the linear space are basis solutions of the linear differential equation known as the time dependent Schrödinger equation, TDSE.

So, we solve the TDSE. We do this by separating the TDSE as we did with the Newtonian wave equation. One part of the separated TDSE is the time independent Schrödinger equation, TISE. The solutions of the TISE are basis functions, ϕ_i, of a linear space (standing wave solutions). The linear space is then all the linear sums of these basis functions and is of the form:

$$\Psi = c_1\varphi_1 + c_2\varphi_2 + c_2\varphi_2 + \ldots$$
$$c_i \in \mathbb{C}$$

(8.1)

The TISE is an energy operator equation, also known as an eigenvalue equation. This means that its basis solutions are energy eigenfunctions. Associated with each eigenfunction is a special number called an eigenvalue. We saw this with the Newtonian wave equation in which each standing wave solution (basis solution) of the Newtonian wave equation is associated with a special number, $k_n = \dfrac{n\pi}{L}$.

In quantum mechanics, the eigenvalues of a linear operator are the possible results of measuring the physical property (momentum, energy,...) associated with the linear operator. In the TISE, the operator is the energy operator, and so the eigenvalues are the energies of the standing wave (eigenfunction) solutions of the TISE.

There are linear operators in any linear space.

8.1   OPERATORS

An operator, $\widehat{A}$, acts upon a mathematical function, ψ, to produce another function. A linear operator, $\widehat{A}$, is an operator such that:

$$\widehat{A}\left(a_1\psi + a_2\phi\right) = a_1\widehat{A}\left(\psi\right) + a_2\widehat{A}\left(\phi\right) \quad : \quad a_i \in \mathbb{C} \qquad (8.2)$$

Actually a_i can be an element of any division algebra. An example of a non-linear operator is exponentiation:

$$\widehat{\exp}\left(a_1\psi + a_2\phi\right) \neq a_1\widehat{\exp}(\psi) + a_2\widehat{\exp}(\phi) \qquad (8.3)$$

Alternatively, given a complete set of orthonormal basis vectors, any linear operator can be written as a matrix operator ($n \times n$ matrix) that acts upon a vector, ($n \times 1$ column matrix) to produce another vector.

$$\begin{bmatrix} a & b \\ -b & a \end{bmatrix}\begin{bmatrix} c \\ d \end{bmatrix} = \begin{bmatrix} ac + bd \\ ad - bc \end{bmatrix} \qquad (8.4)$$

An example of a linear operator acting on a function is the differentiation operator that acts upon a function to produce its differential:

$$\widehat{\frac{\partial}{\partial x}}(x^2) = 2x \qquad (8.5)$$

Note that we signify the operator by putting a carat (little hat) above it. The differentiation operator is $\widehat{\dfrac{\partial}{\partial x}}$. The input function (operator argument) is x^2. The output function is $2x$.

Another example is the multiply by x operator that acts upon an input function to produce the product of that function with x:

$$\hat{x}\left(3x^{\frac{5}{2}} + 7\right) = 3x^{\frac{7}{2}} + 7x \tag{8.6}$$

The multiply by x operator is $\hat{x}$.

Aside: The multiply by x operator is the operator that corresponds to the Newtonian dynamic variable that is x-position.

Other examples include the zero operator that converts all input functions into the zero function:

$$\hat{0}(3x^3 + 9) = 0 \tag{8.7}$$

The identity operator that converts all input functions into themselves:

$$\hat{1}\left(e^{5x} - \sin(x^3)\right) = e^{5x} - \sin(x^3) \tag{8.8}$$

And the multiply by $i\hbar$ operator. Note: $i = \sqrt{-1}$:

$$\widehat{i\hbar}\left(e^{5x} - \sin(x^3)\right) = i\hbar\left(e^{5x} - \sin(x^3)\right) \tag{8.9}$$

To reiterate, an operator is something that acts upon a function to produce another function. Alternatively, a matrix operator acts upon a vector to produce another vector:

$$\begin{bmatrix} 1 & 0 & 3 & 2 \\ 0 & 1 & 2 & 3 \\ 3 & 2 & 1 & 0 \\ 2 & 3 & 0 & 1 \end{bmatrix} \begin{bmatrix} a \\ b \\ c \\ d \end{bmatrix} = \begin{bmatrix} a + 3c + 2d \\ b + 2c + 3d \\ 3a + 2b + c \\ 2a + 3b + d \end{bmatrix} \tag{8.10}$$

We see that the concept of an operator acting upon a function to produce another function is mirrored by a matrix acting upon a vector to produce another vector. In quantum mechanics, we use both these formulations. This dual formulation of quantum mechanics can be confusing because it leads to operators being two types of things - square matrices and functionals. These two types of things act respectively upon vectors and functions.

8.2 EIGENFUNCTIONS (EIGENVECTORS)

Eigenfunction means special function. Eigenvector means special vector. Associated with each operator, there are "special functions" ("special vectors") that are special because, when they are acted upon by that operator, they are unchanged except in that they are multiplied by a number. These special functions (special vectors) are called eigenfunctions or eigenvectors. Examples of eigenfunctions of the differentiation operator are $\{e^{2x}, e^{5ix}\}$:

$$\widehat{\frac{\partial}{\partial x}}(e^{2x}) = 2e^{2x} \qquad : \qquad \widehat{\frac{\partial}{\partial x}}(e^{5ix}) = 5ie^{5ix} \qquad (8.11)$$

8.3 EIGENVALUES

Eigenvalue means special value. Associated with each eigenfunction (eigenvector) of an operator is the number by which the operator multiplies the eigenfunction (eigenvector). These special numbers are called eigenvalues. We take e^{2x} to be an eigenfunction of the differentiation operator that is different from e^{5x}. Examples of eigenvalues are the 2 in:

$$\widehat{\frac{\partial}{\partial x}}(e^{2x}) = 2e^{2x} \qquad (8.12)$$

And the 3 in:

$$\widehat{\frac{\partial}{\partial x}}(e^{3x}) = 3e^{3x} \qquad (8.13)$$

Clearly[1], there is one, and only one, eigenvalue for every eigenfunction (eigenvector).

Although the theory of operators allows eigenvalues to be any kind of number, in quantum mechanics all eigenvalues are real numbers. The eigenvalues of an operator are the possible outcomes of measuring some dynamic variable of the system (think the angular

---

[1.] In mathematics, "clearly" means that the statement to which it refers is not only true but can be understood to be true by thinking about it for less than one year, but more than one month.

momentum of an electron in orbit around an atomic nucleus), and so we need them to be real numbers.

Aside: If the value of, say, momentum was a complex number, we would also have to specify the orientation of the axes of the complex plane as part of that number. Since the orientation of axes is arbitrary, the momentum would be arbitrary.

All of the above is mirrored by matrix operators that have special vectors (eigenvectors) which are unchanged when multiplied by the matrix operator other than to be multiplied by a (real in quantum mechanics) number:

$$\begin{bmatrix} 2 & 0 \\ 0 & -1 \end{bmatrix} \begin{bmatrix} 1 \\ 0 \end{bmatrix} = 2 \begin{bmatrix} 1 \\ 0 \end{bmatrix} \tag{8.14}$$

8.4   MATRICES AND EIGENVALUES

Matrices have eigenvalues. We calculate the eigenvalues of a $n \times n$ matrix, A, by solving the equation:

$$\det(A - \lambda I) = 0 \tag{8.15}$$

Wherein I is the $n \times n$ identity matrix and λ are numbers that are the n solutions of this n^{th} power equation; these numbers are called the eigenvalues of the matrix. If we were to diagonalize the $n \times n$ matrix with a symmetry transformation (change of basis), then the n elements on the leading diagonal would be the eigenvalues of the matrix – the other elements would all be zero, of course. If any of the eigenvalues of a matrix, λ, are equal, we say that these eigenvalues are degenerate. We say the same when dealing with eigenvalues in quantum mechanics.

8.5   HERMITIAN OPERATORS

There are particular types of operators that always have real eigenvalues. These particular operators are called Hermitian operators, and they are the type of operators which are used in quantum mechanics. A Hermitian operator is an operator that is both linear and self adjoint. We will explain self adjointness later. By definition, a functional linear operator, $\widehat{A}$, is Hermitian if:

$$\int_{-\infty}^{\infty} dx \; \phi^* \widehat{A}(\psi) = \int_{-\infty}^{\infty} dx \; \psi \left(\widehat{A}(\phi)\right)^* \qquad (8.16)$$

This automatically ensures the operator, $\widehat{A}$, will have real eigenvalues because:

$$\widehat{A}\phi_n = a_n\phi_n \Rightarrow \widehat{A^*}\phi_n^* = a_n^* \phi_n^*$$
$$\int dx \; \phi_n^* \widehat{A}\phi_n = a_n \int dx \; \phi_n^* \phi_n$$
$$\int dx \; \phi_n \widehat{A^*}\phi_n^* = a_n^* \int dx \; \phi_n \phi_n^* \qquad (8.17)$$
$$a_n = a_n^*$$

The differentiation operator, $\dfrac{\partial}{\partial x}$, is not an Hermitian operator, but the momentum operator, $\widehat{p} = -i\hbar\dfrac{\partial}{\partial x}$, is an Hermitian operator. For example, we have:

$$\int_{-\infty}^{\infty} dx \; \phi^* \widehat{p}(\psi) = -i\hbar \int_{-\infty}^{\infty} dx \; \phi^* \frac{\partial \psi}{\partial x}$$
$$= -i\hbar \left[\phi^* \psi \right]_{-\infty}^{\infty} - (-i\hbar) \int_{-\infty}^{\infty} dx \; \psi \frac{\partial \phi^*}{\partial x} \qquad (8.18)$$

We assume that the wavefunction, $\{\psi, \phi\}$ are square integrable functions. This means that their values at $\pm\infty = 0$, and so the first term is zero:

$$\int_{-\infty}^{\infty} dx \; \phi^* \widehat{p}(\psi) = (i\hbar) \int_{-\infty}^{\infty} dx \; \psi \frac{\partial \phi^*}{\partial x}$$
$$= \int_{-\infty}^{\infty} dx \; \psi i\hbar \frac{\partial \phi^*}{\partial x} \qquad (8.19)$$
$$= \int_{-\infty}^{\infty} dx \; \psi \left(\widehat{p}(\phi)\right)^*$$

It is also true that the x-position operator, $\hat{x}$, is Hermitian. It thus follows that the Hamiltonian operator[2], $\widehat{H}$, (the energy operator), being a multiple and a sum of the momentum and position operators, is also a Hermitian operator.

The (non-degenerate) eigenfunctions of a Hermitian operator are orthogonal; that is:

$$\int dx \; \phi_m^{\circ}\phi_n = \delta_{mn}$$

$$\delta_{mn} = 0 \text{ if } m \neq n \quad \& \quad \delta_{mn} = 1 \text{ if } m = n$$

(8.20)

We are going to use the eigenfunctions of Hermitian operators to form the basis elements (basis functions) of a linear space. We would not be able to do this if they were not orthogonal.

8.6 HERMITIAN MATRICES

A matrix operator is Hermitian if it is of the form that it is equal to its conjugate transpose. The conjugate transpose is called the adjoint. We form the conjugate transpose of a matrix by:

i. Conjugate every complex number in the matrix

ii. Transpose the matrix (swap a_{ij} for a_{ji})

An example of a non-Hermitian matrix is:

$$\begin{bmatrix} a & b+ic \\ d+ie & a \end{bmatrix}^{\circ T} = \begin{bmatrix} a & b-ic \\ d-ie & a \end{bmatrix}^{T} = \begin{bmatrix} a & d-ie \\ b-ic & a \end{bmatrix}$$

(8.21)

An example of a Hermitian matrix is:

$$\begin{bmatrix} a & b+ic \\ b-ic & a \end{bmatrix}^{\circ T} = \begin{bmatrix} a & b-ic \\ b+ic & a \end{bmatrix}^{T} = \begin{bmatrix} a & b+ic \\ b-ic & a \end{bmatrix}$$

(8.22)

[2.] We have mentioned the Hamiltonian operator earlier in the book, but not at length. It does not matter if the reader has forgotten it because we will meet it again later.

If a matrix operator is equal to its own adjoint (conjugate transpose), we say that it is self adjoint. We often write the adjoint matrix with a superscripted dagger:

$$\begin{bmatrix} a & b+ic \\ d+ie & a \end{bmatrix}^{\dagger} = \begin{bmatrix} a & d-ie \\ b-ic & a \end{bmatrix} \tag{8.23}$$

8.7 UNITARY MATRICES

A unitary matrix is one where:

$$UU^{\dagger} = U^{\dagger}U = I$$
$$U^{-1} = U^{\dagger} \tag{8.24}$$

Wherein $U^{\dagger}$ is the adjoint (conjugate transpose) of U. Within a unitary matrix, the rows form an orthonormal set of basis vectors and the columns also form an orthonormal set of basis vectors. The eigenvalues, λ_i, of a unitary matrix are such that $|\lambda_i|^2 = 1$; they are of unit length.

Aside: A general 2×2 unitary matrix is of the form:

$$\begin{bmatrix} c_1 & c_2 \\ -c_2^{\circ} & c_1^{\circ} \end{bmatrix} \; : \; |c_1|^2 + |c_2|^2 = 1 \tag{8.25}$$

This is equivalent to:

$$\begin{bmatrix} a & b & c & d \\ -b & a & -d & c \\ -c & d & a & -b \\ -d & -c & b & a \end{bmatrix} \; : \; a^2 + b^2 + c^2 + d^2 = 1 \tag{8.26}$$

Which is a quaternion of unit length. The quaternion rotation matrix is isomorphic with the Lie group $SU(2)$ which the reader will meet in later studies.

Suppose I do an experiment with apparatus that is pointing north. If I do the same experiment with the same apparatus pointing west, will I get the same result? The directions north or west are no more than the arbitrary whims of humankind, and we would be amazed if our whims were able to affect the outcome of an experiment. We take the view that the physical laws of the universe are independent of the direction in which the experimental apparatus points. This is a view very much supported both theoretically and observationally. If the laws of physics are unchanged by a change of viewpoint, we say that the laws of physics are symmetrical with respect to the change of view.

One of the results of experiments is the inner product of two wavefunctions which we associate with the probability of a particular result being seen from an experiment.

$$P \propto \langle \phi_i, \psi \rangle^2 \tag{8.27}$$

It is a mathematical fact that, for a unitary linear operator, U:

$$\langle \phi_i, \psi \rangle^2 = \langle U(\phi_i), U(\psi) \rangle^2 \tag{8.28}$$

Now, the operator U is mapping vectors within a linear space to other vectors within that linear space, and it does it in such a way that the inner product of the two vectors is the same (invariant). What is really happening is no more than a change of basis of the linear space. The change of basis is a rotation in unitary space.

Aside: Lie algebra is concerned with symmetries corresponding to operators that can be written as:

$$U = 1 + i\varepsilon G + O(\varepsilon^2)... \tag{8.29}$$

Applying the condition of unitarity to this operator gives:

$$\left(1 - i\varepsilon G^\dagger + O(\varepsilon^2)...\right)\left(1 + i\varepsilon G^\dagger + O(\varepsilon^2)...\right) = 1 \tag{8.30}$$

Ignoring higher terms, this is:

$$G = G^\dagger \tag{8.31}$$

If we set $\varepsilon = \dfrac{\theta}{n}$ and apply the operator n times, as $n \to \infty$, we get:

$$U(\theta) = \left(1 + i\frac{\theta}{n}G\right)^n \to e^{i\theta G} \tag{8.32}$$

This is a rotation in the complex plane.

8.8 SIMILARITY TRANSFORMATIONS

A similarity transformation is a transformation that changes the basis in which the matrix, A, is written. It is of the form:

$$D = S^{-1}AS \qquad (8.33)$$

Wherein S is an invertible matrix. The matrix D is the same matrix as the matrix A but it is written in a different basis. We note that if S is a unitary matrix, we can replace S^{-1} with $S^{\dagger}$. It is possible to choose a matrix basis such that the eigenvalues of the matrix appear on the leading diagonal of the matrix and all other elements of the matrix are zero. This is called diagonalization of the matrix. It is sometimes useful to diagonalize Hermitian matrices using unitary matrices in the similarity transformation. We then have the Hermitian matrix written in diagonal form and in non-diagonal form; it is the same matrix but written in two different bases. Such a transformation is called a unitary transformation and is written:

$$D = U^{\dagger}AU \qquad (8.34)$$

In due course, the reader will be told that the eigenfunctions of the quantum mechanical momentum operator, $\widehat{P_x}$, are not the same as the eigenfunctions of the quantum mechanical x-position operator, $\hat{x}$, and that this is why we cannot simultaneously know both the momentum and the position of a particle. The two sets of eigenfunctions differ only in that they are written in a different basis – they are the same eigenfunctions written in two different bases. One set of eigenfunctions can be transformed into the other set of eigenfunctions by a unitary similarity transformation. The similarity transformation has to be unitary if the "size" (modulus) of the eigenfunctions is to remain unchanged.

In essence, a unitary transformation is a rotation of the basis of a linear space.

8.9   SYMMETRIC MATRICES ALWAYS HAVE REAL EIGENVALUES

A symmetric matrix is one that is equal to its transpose. This means that the elements of the matrix are such that $a_{rc} = a_{cr}$. We have:

$$\begin{bmatrix} a & 0 & c & d \\ 0 & a & -d & c \\ c & -d & a & 0 \\ d & c & 0 & a \end{bmatrix}^T = \begin{bmatrix} a & 0 & c & d \\ 0 & a & -d & c \\ c & -d & a & 0 \\ d & c & 0 & a \end{bmatrix} \tag{8.35}$$

Wherein the superscript T indicates transposition. This matrix has eigenvectors:

$$\begin{bmatrix} \dfrac{c}{\sqrt{c^2 + d^2}} \\ \dfrac{-d}{\sqrt{c^2 + d^2}} \\ 1 \\ 0 \end{bmatrix}, \begin{bmatrix} \dfrac{d}{\sqrt{c^2 + d^2}} \\ \dfrac{c}{\sqrt{c^2 + d^2}} \\ 0 \\ 1 \end{bmatrix} \tag{8.36}$$

Both with eigenvalue $a + \sqrt{c^2 + d^2}$, and eigenvectors:

$$\begin{bmatrix} \dfrac{-c}{\sqrt{c^2 + d^2}} \\ \dfrac{d}{\sqrt{c^2 + d^2}} \\ 1 \\ 0 \end{bmatrix}, \begin{bmatrix} \dfrac{-d}{\sqrt{c^2 + d^2}} \\ \dfrac{-c}{\sqrt{c^2 + d^2}} \\ 0 \\ 1 \end{bmatrix} \tag{8.37}$$

Both with eigenvalue $a - \sqrt{c^2 + d^2}$. We see that the eigenvalues will always be real.

It is a mathematical fact that symmetric matrices always have real eigenvalues. It is also a mathematical fact that $n \times n$ symmetric matrices with real elements have n mutually orthogonal eigenvectors. The eigenvectors of a quantum mechanical operator are the basis of a linear space in that they are orthogonal and are a complete set. The reader might think that we should have only symmetric matrix operators within quantum mechanics. This makes a lot of sense, but, for historical reasons, in quantum mechanics, this is not done; we use Hermitian matrices to save paper.

8.10 FROM OVERLAP INTEGRAL TO HERMITIAN MATRIX

If a functional Hermitian operator has a finite number of basis eigenfunctions, say n. Given a set of basis vectors, we can define a square $n \times n$ matrix in which the elements of that matrix, M, are given by the inner products of the basis eigenfunctions with each other:

$$M_{RC} = \int_{-\infty}^{\infty} dx\ \phi_R{}^\circ \widehat{A}(\phi_C)$$

$$= \langle \phi_R, \phi_C \rangle$$

(8.38)

Wherein ϕ_i are the eigenfunctions of the operator $\widehat{A}$. Such a matrix is the Hermitian functional operator written as an Hermitian matrix operator. This matrix will be a diagonal matrix (have off-diagonal elements that are all zero) if the matrix is written in the same basis as the eigenfunctions; otherwise, the matrix will not be a diagonal matrix.

8.11 EIGENVALUE EQUATIONS

Equations of the form:

$$\widehat{A}(u_n(x)) = a_n u_n(x)$$

(8.39)

are known as eigenvalue equations. $\widehat{A}$ is an operator. $u_n(x)$ is an eigenfunction, and a_n is the eigenvalue of the particular eigenfunction. An example is given by the momentum operator:

$$\widehat{-i\hbar \frac{\partial}{\partial x}}\left(e^{i\frac{a}{\hbar}x}\right) = -i\hbar\left(i\frac{a}{\hbar}e^{i\frac{a}{\hbar}x}\right) = ae^{i\frac{a}{\hbar}x}$$

(8.40)

Eigenvalue equations are central to quantum mechanics. It is by solving the eigenvalue equations that we get the eigenvalues which are the possible outcomes of measuring the Newtonian dynamic variable (momentum, position, energy, angular momentum,....) associated with the operator. The time independent Schrödinger equation, which is the backbone of much quantum mechanics, is an (energy operator) eigenvalue equation. We also get the eigenfunctions from an eigenvalue equation, and these enable the calculation of the probability of a particular outcome of an event.

All of this is matched in the matrix mechanics representation of quantum mechanics in which an eigenvalue equation is a matrix equation of the form:

$$\begin{bmatrix} \cdot & & \\ & \cdot & \\ & & \cdot \end{bmatrix} \begin{bmatrix} \\ \\ \end{bmatrix} = a_n \begin{bmatrix} \cdot \\ \\ \end{bmatrix} \tag{8.41}$$

We solve an eigenvalue equation by educated guesswork.

8.12 UPON EIGENVALUES

There is a most important physical law which describes the universe. This most important physical law is that the physics of the universe is the same regardless of which co-ordinate system we choose to place over the universe. Electrons have the same electric charge in polar co-ordinates as they have in Cartesian co-ordinates. Electrons have the same electric charge in Cartesian co-ordinates that are oriented at 45° as they have in Cartesian co-ordinates that are oriented at 20°.

It is the very essence of special relativity that physics is the same regardless of the orientation of the space-time co-ordinates. The relative orientation of the space-time co-ordinates of two observers is just the relative velocity of the observers, and physics is the same at all velocities.

This invariance of physics under change of co-ordinates (change of matrix basis) is central to quantum field theory. A central part of

QFT is that physics is invariant under change of co-ordinates in unitary spaces like $\mathbb{C}^2$ and $\mathbb{C}^3$. Rotation of co-ordinates in these unitary spaces is called invariance under $SU(2)$ transformations and invariance under $SU(3)$ transformations respectively.

If we are to formulate a description of the universe, the variables and the equations that relate these variables must be invariant under change of co-ordinates just as the energy of an electron in orbit must be invariant under change of co-ordinates.

Eigenvalues, like the determinant which is their product, are invariant under change of basis (change of co-ordinates). Once we have a matrix operator, its eigenvalues will be the same regardless of which basis we use to write the operator. The eigenvalues are invariants of the system. So, if you want a description of the universe which is linear and independent of co-ordinate system, and the universe certainly seems to be like this, you might expect that eigenvalues would play a part.

8.13   PRODUCTS AND SUMS OF OPERATORS

The "product" of two operators is an operator. This product of operators is "first let one operator act on a function to produce a second function and then let the second operator act upon the second function to produce a third function". Since operators are no more than things that change a function into another function, we see that $\widehat{A}\big(\widehat{B}(f(x))\big)$ is itself an operator.

$$\widehat{A}\big(\widehat{B}(f(x))\big) = \widehat{A}\big(g(x)\big) = h(x) \qquad (8.42)$$

Clearly, the "sum" of two operators is an operator.

$$\big(\widehat{A}+\widehat{B}\big)(f(x)) = \widehat{A}\big(f(x)\big) + \widehat{B}\big(f(x)\big) = g(x) + h(x) = j(x) \quad (8.43)$$

Alternatively, a product of matrix operators starts with a vector and first converts it to a second vector and then converts the second vector to a third vector.

8.14 THE COMMUTATOR OF TWO OPERATORS

It is a simple mathematical fact, not based on any assumptions, that the order in which different operators act upon a function affects the outcome. Consider the two operators:

$$\widehat{A}(f(x)) = \frac{\widehat{\partial}}{\partial x} f(x), \qquad \widehat{B}(f(x)) = \widehat{x}.f(x) \qquad (8.44)$$

The $\widehat{A}$ operator is differentiate a function with respect to x, and the $\widehat{B}$ operator is multiply the function by x. The order in which these operators act upon the function, $f(x)$, matters. Consider the case $f(x) = x^2 + 7$:

$$\begin{aligned} \widehat{A}\big(\widehat{B}(f(x))\big) &= \widehat{A}\big(\widehat{B}(x^2 + 7)\big) = \widehat{A}(x^3 + 7x) = 3x^2 + 7 \\ \widehat{B}\big(\widehat{A}(f(x))\big) &= \widehat{B}\big(\widehat{A}(x^2 + 7)\big) = \widehat{B}(2x) \qquad = 2x^2 \end{aligned} \qquad (8.45)$$

We see that, by changing the order in which the operators act upon the function, we produce different results.

The commutator of two operators is the difference between the alternately ordered products of the operators. The commutator of the two operators is also an operator. The commutator is written as a square bracket:

$$\big[\widehat{A},\widehat{B}\big]f(x) = \widehat{A}\big(\widehat{B}(f(x))\big) - \widehat{B}\big(\widehat{A}(f(x))\big) \qquad (8.46)$$

In the above example, this would be:

$$\big[\widehat{A},\widehat{B}\big](x^2 + 7) = (3x^2 + 7) - (2x^2) = x^2 + 7 \qquad (8.47)$$

Which is the function with which we started. This is not a coincidence; it is true in general for the two operators above. It is not true for most operators. In the above case, since the commutator operator is the identity operator 1, we have:

$$\left[\frac{\widehat{\partial}}{\partial x}, \widehat{x}\right] = \widehat{1} \qquad (8.48)$$

Within which, we have placed carats over each of the elements because they are all operators. With a little thought, the reader will realizethat swapping the order of the operators within the bracket reverses the sign of the commutator operator:

$$\left[\hat{x}, \frac{\widehat{\partial}}{\partial x}\right] = -\hat{1} \tag{8.49}$$

We refer to expressions such as the immediately above as commutation relations.

8.15 MATRIX COMMUTATORS

The commutation relations of quantum mechanical operators can be seen more clearly when we use matrix operators:

$$\begin{bmatrix} 0 & 1 \\ 1 & 0 \end{bmatrix} \begin{bmatrix} 0 & -i \\ i & 0 \end{bmatrix} - \begin{bmatrix} 0 & -i \\ i & 0 \end{bmatrix} \begin{bmatrix} 0 & 1 \\ 1 & 0 \end{bmatrix} = \begin{bmatrix} i & 0 \\ 0 & -i \end{bmatrix} - \begin{bmatrix} -i & 0 \\ 0 & i \end{bmatrix}$$
$$= \begin{bmatrix} 2i & 0 \\ 0 & -2i \end{bmatrix} \neq 0 \tag{8.50}$$

8.16 COMPATIBLE OPERATORS

Imagine two operators, $\{\widehat{A}, \widehat{B}\}$, which share the same eigenfunctions, ϕ_n. We have:

$$\widehat{A}(\phi_n) = a_n \phi_n \quad \& \quad \widehat{B}(\phi_n) = b_n \phi_n \tag{8.51}$$

And:

$$\widehat{B}\left(\widehat{A}(\phi_n)\right) = b_n a_n \phi_n \quad \& \quad \widehat{A}\left(\widehat{B}(\phi_n)\right) = a_n b_n \phi_n \tag{8.52}$$

Which implies:

$$\left[\widehat{A}, \widehat{B}\right]\phi_n = 0 \tag{8.53}$$

In other words, if two operators share the same eigenfunctions, they commute. Conversely, if two operators commute, they have the same eigenfunctions.

Commuting operators need not associate the same eigenvalues with any particular shared eigenfunction. The momentum operator,

$\hat{p}$, and the zero potential energy operator, $\widehat{H}$, have the same set of eigenfunctions, but they have different sets of eigenvalues.

Commuting operators are said to be compatible. Conversely, if two operators do not have the same eigenfunctions, they do not commute and are said to be incompatible. Since a state can be described by only one eigenfunction (one standing wave) at a time, we cannot simultaneously know the eigenvalues of two incompatible operators. The x-position operator and the x-momentum operator are incompatible operators; they do not commute, and we cannot simultaneously know both the x-momentum and the x-position of a particle.

8.17   INCOMPATIBLE EIGENFUNCTIONS AND UNITARY TRANSFORMATIONS

When we deal with intrinsic spin, we will meet the spin operators. The spin operators are the Pauli matrices multiplied by $\frac{\hbar}{2}$. The spin operators are:

$$S_x = \frac{\hbar}{2}\begin{bmatrix} 0 & 1 \\ 1 & 0 \end{bmatrix}, \quad S_y = \frac{\hbar}{2}\begin{bmatrix} 0 & -i \\ i & 0 \end{bmatrix}, \quad S_z = \frac{\hbar}{2}\begin{bmatrix} 1 & 0 \\ 0 & -1 \end{bmatrix} \tag{8.54}$$

The spin operators are the same matrix in different bases. We have the unitary matrix:

$$U_{x \to z} = \frac{1}{\sqrt{2}}\begin{bmatrix} 1 & -1 \\ 1 & 1 \end{bmatrix}, \quad U_{x \to z}^{\dagger} = \frac{1}{\sqrt{2}}\begin{bmatrix} 1 & 1 \\ -1 & 1 \end{bmatrix}$$

$$U_{x \to z}^{\dagger} U_{x \to z} = \begin{bmatrix} 1 & 0 \\ 0 & 1 \end{bmatrix} \tag{8.55}$$

We do the unitary transformation on S_x:

$$U_{x \to z}^{\dagger} S_x U_{x \to z} = \frac{1}{\sqrt{2}}\begin{bmatrix} 1 & 1 \\ -1 & 1 \end{bmatrix}\frac{\hbar}{2}\begin{bmatrix} 0 & 1 \\ 1 & 0 \end{bmatrix}\frac{1}{\sqrt{2}}\begin{bmatrix} 1 & -1 \\ 1 & 1 \end{bmatrix}$$

$$= \frac{\hbar}{2}\begin{bmatrix} 1 & 0 \\ 0 & -1 \end{bmatrix} = S_z \tag{8.56}$$

We see that S_x is the same operator as S_z except that they are written in different bases. The same applies to the S_y operator, but we need a different unitary transformation. If $\{S_x, S_z\}$ are the same operator, then they have the same eigenvectors except that the eigenvectors are written in different bases. This is true in general of incompatible operators and their eigenfunctions (eigenvectors).

That nature uses three different bases for the spin operators is most strange. We would expect that one basis would be sufficient. It seems that nature is working with three different, but only by basis, types of space-time. Though familiarity, physicists have come to accept non-commutative operators and the impossibility of simultaneously observing incompatible variables as normal. In doing so, they sweep under the carpet this most mysterious aspect of quantum mechanics.

8.18 SUMMARY

We are interested in operators that multiply a function (vector) by a real number. We call these operators Hermitian operators. We are interested in solving eigenvalue equations. We are interested in the commutation relations of operators. Incompatible operators are the same operator in different bases.

8.19 BRAS AND KETS (ADDENDUM)

In the Dirac (bra & ket) notation, we write an operator with a carat above it, and so an eigenvalue equation is written as:

$$\widehat{A}|\psi\rangle = a|\phi\rangle \qquad (8.57)$$

We have:

$$\widehat{A+B}|\psi\rangle = \widehat{A}|\psi\rangle + \widehat{B}|\psi\rangle$$
$$\widehat{AB}|\psi\rangle = \widehat{A}\left(\widehat{B}|\psi\rangle\right) \qquad (8.58)$$

Operators are to the left of kets but to the write of bras:

$$\widehat{A}|\psi\rangle \quad \& \quad \langle\psi|\widehat{A} \tag{8.59}$$

So we have:

$$\langle\phi|\widehat{A}|\varphi\rangle \equiv \langle\phi\widehat{A}\varphi\rangle \equiv \begin{bmatrix} \phi_1 & \phi_2 \end{bmatrix} \begin{bmatrix} A_{11} & A_{12} \\ A_{21} & A_{22} \end{bmatrix} \begin{bmatrix} \varphi_1 \\ \varphi_2 \end{bmatrix} \tag{8.60}$$

If $\{\phi, \varphi\}$ are basis vectors of the space, then the operator, $\widehat{A}$, is given by:

$$\widehat{A} = \begin{bmatrix} \langle\phi|\widehat{A}|\phi\rangle & \langle\phi|\widehat{A}|\varphi\rangle \\ \langle\varphi|\widehat{A}|\phi\rangle & \langle\varphi|\widehat{A}|\varphi\rangle \end{bmatrix} \tag{8.61}$$

The adjoint of the Dirac expression:

$$c\langle\psi|\widehat{A}\widehat{B}|\phi\rangle \tag{8.62}$$

Is the expression:

$$c^\circ\langle\phi|\widehat{B}^\dagger\widehat{A}^\dagger|\psi\rangle \tag{8.63}$$

An Hermitian operator is one where:

$$\widehat{A} = \widehat{A}^\dagger$$

$$\langle\psi|\widehat{A}|\phi\rangle = \left(\langle\phi|\widehat{A}|\psi\rangle\right)^\circ \tag{8.64}$$

EXERCISES

1. Prove that if two operators commute, they have the same eigenfunctions.

2. Calculate the commutator of the two symmetric matrices:

$$\begin{bmatrix} 0 & 0 & 1 & 0 \\ 0 & 0 & 0 & 1 \\ 1 & 0 & 0 & 0 \\ 0 & 1 & 0 & 0 \end{bmatrix} \quad \& \quad \begin{bmatrix} 0 & 0 & 0 & 1 \\ 0 & 0 & -1 & 0 \\ 0 & -1 & 0 & 0 \\ 1 & 0 & 0 & 0 \end{bmatrix} \tag{8.65}$$

Is the product of the above two matrices a symmetric matrix?

3. Symmetric matrices are associated with rotations in space-time (special relativity) – just take the exponential of the matrix. Anti-symmetric matrices are associated with rotations in space. Looking at the answer to 1, is the commutator of two space-time rotations a spatial rotation?

4. Is the operator $\widehat{\dfrac{\partial}{\partial x}}$ acting upon $\psi(x)$ Hermitian?

5. The "Hadamard gate" is the matrix:

$$H = \frac{1}{\sqrt{2}}\begin{bmatrix} 1 & 1 \\ 1 & -1 \end{bmatrix} \tag{8.66}$$

Is this matrix Hermitian? Is this matrix unitary? What is the modulus of the determinant of this matrix?

6. Confirm the operator equation:

$$\overline{\left(\frac{\partial}{\partial x}+x\right)}\left(\left(\widehat{\frac{\partial}{\partial x}}-\hat{x}\right)\right)=\widehat{\frac{\partial^2}{\partial x^2}}-\widehat{x^2}-\hat{1} \tag{8.67}$$

7. Evaluate the operator:

$$\overline{\left(\frac{\partial}{\partial x}-x\right)}\left(\left(\widehat{\frac{\partial}{\partial x}}+\hat{x}\right)\right) \tag{8.68}$$

8. Is $e^{-\frac{x^2}{2}}$ an eigenfunction of the operator $\overline{\dfrac{\partial^2}{\partial x^2}}-x^2$?

AUTHOR'S NOTE

The reader has now just finished the two most difficult chapters of this book. From now on, it is downhill. If the reader has not yet done so, she is urged to re-read the last two chapters.

9

THE PLACE OF OPERATORS IN QUANTUM MECHANICS

Quantum mechanics is formulated in two representations. The wave mechanics representation is built upon linear operators, eigenfunctions, and eigenvalues. The matrix mechanics representation is built upon matrix operators, eigenvectors, and eigenvalues. Matrix operators are naturally linear operators – matrix algebra is linear algebra. It is normal to refer to both types of operator as just "operators."

9.1 CORRESPONDENCE PRINCIPLE

i. For every dynamic variable in Newtonian mechanics, there is a corresponding operator in quantum mechanics.

ii. The relations between Newtonian dynamic variables are duplicated within quantum mechanics as relations between operators.

iii. There are commutation relations between operators that do not exist between Newtonian variables. These commutation relations are an essential difference between quantum mechanics and Newtonian mechanics.

9.2 OPERATORS

In quantum mechanics, every Newtonian dynamic variable (except time), like energy, or momentum, or position, or angular momentum, or..., is replaced by a single corresponding linear operator known as the energy operator, or the momentum operator, or the position operator, or.... For example, the x-momentum variable of Newtonian mechanics, p, corresponds to the x-momentum operator, $\widehat{p_x}$, of quantum mechanics. This x-momentum operator is:

$$\widehat{p_x} = -i\hbar \frac{\partial}{\partial x} \qquad (9.1)$$

Aside: We sometimes see the momentum operator written as $\widehat{\pi}$ instead of as $\hat{p}$. This is common in QFT.

This association of Newtonian dynamic variables with linear operators is a postulate of quantum mechanics; it cannot be proven mathematically. Indeed, we would be happier if we could do without it and could get our operators from elsewhere (symmetry considerations perhaps).

Within Newtonian mechanics, there are relations between dynamic variables such as the relation between kinetic energy, E, and momentum, p, in a zero potential:

$$E = \frac{1}{2}mv^2 = \frac{p^2}{2m} \qquad (9.2)$$

These relations are duplicated by the operators in quantum mechanics, and so, from the momentum operator, we are able to deduce the form of the zero potential energy operator (written as either $\widehat{E}$ or $\widehat{H}$).

$$\begin{aligned} \widehat{E} &= \frac{1}{2m}\hat{p}^2 \\ &= \frac{1}{2m}\left(-i\hbar\frac{\partial}{\partial x}\left(-i\hbar\frac{\partial}{\partial x}\right)\right) \\ &= -\frac{\hbar^2}{2m}\frac{\partial^2}{\partial x^2} \end{aligned} \qquad (9.3)$$

Operators vary from one physical system to another physical system. For example, the non-zero potential energy operator includes a potential term, $V(t, x)$:

$$\widehat{E} = -\frac{\hbar^2}{2m}\frac{\partial^2}{\partial x^2} + V(t,x) \qquad (9.4)$$

This potential might be zero (a free particle) or it might be a spherical potential or it might be something else. We see that the energy operator varies from one potential to another. Since the amount of energy varies from one potential to another, this ought not to surprise the alert reader.

9.3   THE RADIOACTIVE DECAY OF TRITIUM

Tritium is an isotope of hydrogen with two neutrons and one proton. It is unstable and decays with a half-life of 12.3 days. Tritium decays into an isotope of helium with two protons and one neutron. Effectively, a neutron changes into a proton. When tritium decays, it emits an electron and an antineutrino with total energy release of 18.6 kilo electron volts.

There is a single electron orbiting the tritium nucleus. When the tritium nucleus decays, the potential of the atom changes due to the doubling of the number of protons and the Hamiltonian (energy operator) describing the electron must change.

9.4   EIGENFUNCTIONS (EIGENVECTORS)

Associated with each quantum mechanical operator, there is a set, which might be of infinite number, of eigenfunctions (eigenvectors) known as energy eigenfunctions which are eigenfunctions of the energy operator, or momentum eigenfunctions, or position eigenfunctions, or... For example, the x-momentum eigenfunctions are of the form:

$$\phi = Ae^{i\frac{p_x}{\hbar}x} \qquad (9.5)$$

Where A is a constant that we can adjust to normalize the wave-function that is a superposition (linear sum) of these eigenfunctions and $p_x \in \mathbb{R}$ is the associated eigenvalue which is the value of the momentum associated with the particular state described by the particular eigenfunction. The eigenfunctions are complex (the equivalent eigenvectors have complex components).

9.5   THE RADIOACTIVE DECAY OF TRITIUM REVISITED

The two Hamiltonians involved in tritium decay do not have the same eigenfunctions and so the electron must "change its wavefunction" from a superposition of the eigenfunctions of the old Hamiltonian to a superposition of the eigenfunctions of the new Hamiltonian. Different potentials have different energy operators and so different energy eigenfunctions.

9.6   COMMUTATION OF OPERATORS

If two particular operators commute, their respective sets of eigenfunctions (linear spaces of eigenfunctions) are the same; they have the same eigenfunctions written in the same basis. This does not mean that the have the same eigenvalues, just the same eigenfunctions.

If two particular operators do not commute, their eigenfunctions are not the same; they do not have the same eigenfunctions, but this will be due only to the basis of the linear spaces being different.

The set of eigenfunctions (eigenvectors) of any operator will vary from one physical system to another. For example, in a zero potential, the energy operator and the x-momentum operator commute and so share the same linear space of eigenfunctions written in the same basis. In a non-zero potential (non-uniform potential), the energy operator and the x-momentum operator do not commute and so do not share the same linear space of eigenfunctions (eigenvectors). The difference between the two linear spaces is only that they are written in different bases.

9.7   EIGENVALUES

With each eigenfunction (eigenvector) of a particular operator, there is an associated eigenvalue. The set of these eigenvalues are known as the energy eigenvalues (the eigenvalues of the eigenfunctions of the energy operator), or the momentum eigenvalues, or the position eigenvalues, or.... In quantum mechanics, these eigenvalues are the only possible values of the associated Newtonian dynamic variable. In Newtonian dynamics, the dynamic variables like energy, or momentum, or position, or ... are all continuous variables; they can take any real value. In quantum mechanics, these dynamic variables can take only values or that are eigenvalues of the associated operator.

Sometimes, say an electron orbiting an atomic nucleus, the eigenvalues of an operator are a discreet set of values. This is a consequence of boundary conditions associated with a particle (which is a wave) bound within a potential well – think Newtonian wave equation and standing waves. In such cases, the only possible values of the Newtonian variable associated with that operator are the discreet values of the eigenvalues. The angular momenta of an orbiting electron are an example of this discreetness of possible values.

Other times, say a freely moving electron, the eigenvalues of an operator are a continuous set of values – just like Newtonian dynamic variables.

EXERCISES

1. Verify that $[A, B] = - [B, A]$?

2. Verify that $[A + B, C] = [A, C] + [B, C]$?

3. Is the matrix:

$$R = \begin{bmatrix} \cos\theta & \sin\theta \\ -\sin\theta & \cos\theta \end{bmatrix} \qquad (9.6)$$

a unitary matrix? Is it Hermitian?

4. Rotation in anti-quaternion space is encapsulated in the anti-quaternion rotation matrix:

$$\begin{bmatrix} \cos(\lambda) & \dfrac{b}{\lambda}\sin(\lambda) & \dfrac{c}{\lambda}\sin(\lambda) & \dfrac{d}{\lambda}\sin(\lambda) \\[2ex] -\dfrac{b}{\lambda}\sin(\lambda) & \cos(\lambda) & \dfrac{d}{\lambda}\sin(\lambda) & -\dfrac{c}{\lambda}\sin(\lambda) \\[2ex] -\dfrac{c}{\lambda}\sin(\lambda) & -\dfrac{d}{\lambda}\sin(\lambda) & \cos(\lambda) & \dfrac{b}{\lambda}\sin(\lambda) \\[2ex] -\dfrac{d}{\lambda}\sin(\lambda) & \dfrac{c}{\lambda}\sin(\lambda) & -\dfrac{b}{\lambda}\sin(\lambda) & \cos(\lambda) \end{bmatrix} \qquad (9.7)$$

$$\lambda = \sqrt{b^2 + c^2 + d^2}$$

Is this a unitary matrix?

5. Verify that $[A, B]^{\dagger} = -[A^{\dagger}, B^{\dagger}]$?

10

COMMUTATION RELATIONS IN QUANTUM MECHANICS

If two operators share the same set of eigenfunctions, then the two operators commute with each other. We have previously shown a proof of this. The converse is also true; if two operators commute, they have the same set of eigenfunctions. For non-degenerate eigenfunctions, we have:

$$\left[\widehat{A}, \widehat{B}\right] = 0$$
$$\widehat{A}\left(\widehat{B}(\phi_n)\right) = \widehat{B}\left(\widehat{A}(\psi_m)\right)$$
$$\widehat{A}(b_n\phi_n) = \widehat{B}\left(a_m\psi_m\right)$$
$$a_k b_n \phi_n = b_l a_m \psi_m$$
$$\phi_n = C\psi_m$$

(10.1)

At most, the eigenfunctions of two commuting operators can differ by a real number multiple.

Prior to observation, we view a physical system as being in a superposition (linear sum) of all possible states. When the superposition (wavefunction) of a physical system is observed, it collapses into an eigenstate which is an eigenfunction of a particular operator.

If this eigenfunction is also the eigenfunction of a different operator, then there are two eigenvalues associated with the eigenfunction – one for each operator – and the physical system is in a state with these eigenvalues as values of the two dynamic variables. Thus, we are able to know simultaneously, the value of both the dynamic variables associated with the two operators.

If two operators do not commute with each other, then they will not have the same eigenfunctions. When the wavefunction (superposition of eigenfunctions) collapses, it will collapse into an eigenstate of only one of the operators. Associated with this eigenstate will be an eigenvalue that is the value of the dynamic variable associated with this first operator. However, since the eigenstate is not an eigenfunction of the other operator, there will not be an eigenvalue of this second operator, and so there will be no value associated with the dynamic variable of the second operator. This means that it is not possible to measure the dynamic variables associated with the two operators simultaneously.

The "choice" of collapse into a particular eigenstate is a "choice" of collapse into a particular basis.

Aside: The non-commutativity of operators is often said to be the fundamental difference between quantum mechanics and classical physics. However, we find the commutator of covariant derivatives within general relativity. The Riemann tensor measures that bit of the commutator of covariant derivatives which is proportional to a vector field. If we take the view that classical physics does not include general relativity, then non-commutativity of operators is a fundamental difference between quantum mechanics and classical physics.

In a later chapter, we will derive the eigenvalues of angular momentum from nothing more than the commutation relations of the angular momentum operators. In this sense, we might view commutation relations as being both basic and central to reality.

10.1 THE COMMUTATOR OF MOMENTUM AND POSITION

The commutator of position and momentum is a multiple of $\hbar$. We have:

$$
\begin{aligned}
\left[\widehat{p_x}, x\right]\psi &= -i\hbar\frac{\partial}{\partial x}(x\psi) + xi\hbar\frac{\partial\psi}{\partial x} \\
&= -i\hbar\left(x\hbar\frac{\partial\psi}{\partial x} + \psi - x\frac{\partial\psi}{\partial x}\right) \\
&= -i\hbar\psi
\end{aligned}
\tag{10.2}
$$

It turns out that all commutators within quantum mechanics are either zero or are multiples of $\hbar$.

10.2 DISTURBANCE AND THE COMMUTATOR OF OPERATORS

We learned above that the commutator of two operators is not necessarily zero. There is a physical explanation that goes with this observation. It is a matter of opinion whether or not this explanation is convincing. We leave it to the reader to form their own opinion.

Suppose we first measure the position of an electron and, after that, we measure the momentum of that electron. The value of the measurement of the position will be an eigenvalue of the position operator. We will then measure the momentum of the electron with that particular eigenvalue of position and get an eigenvalue of the momentum operator.

Now suppose we do the measurements in the opposite order. First we measure the momentum of the electron and get an eigenvalue of the momentum operator. It is not necessarily true that this particular eigenvalue of the momentum operator will be the same momentum eigenvalue as the one we got when we measured the momentum after we had first measured the position of the electron. If we next measure the position of the electron, it is not necessarily true that we will get the same position eigenvalue as we got when we first measured the position of the electron before we measured the momentum of the electron. Indeed, experiments seem to show that we do not get the same measurements of the position and momentum of an electron if we reverse the order in time of the measurements.

The standard explanation of this is that measuring the position of something as small as an electron necessarily disturbs that electron. To measure the electron's position, we need to hit the electron with a photon. This hitting the electron with a photon will change the momentum of the electron and thus change the particular momentum eigenfunctions associated with that electron – if we've changed the system, we've changed the eigenfunctions that describe the system. The new momentum eigenfunctions of the electron will have momentum eigenvalues that are different from the old momentum eigenfunctions. If, after measuring the position of an electron, we then measure its momentum, we ought to expect to get a different result from that which we would have got if we had measured the electron's momentum before measuring the electron's position because we are measuring the momentum of an electron with a changed momentum and thus a changed set of momentum eigenfunctions and therefore with a changed set of momentum eigenvalues.

Of course, if we measure the momentum of a cricket ball by hitting the cricket ball with a photon, then we would cause negligible disturbance to the position of the cricket ball, and so it would seem that we can measure both the position and the momentum of a cricket ball in either order and get the same answers. This is not exactly true, but it is so near to being exactly true that it is exactly true[1].

10.3   UNCERTAINTY IN QUANTUM MECHANICS

If a particle is a wave in space, there is an inherent uncertainty in the position of that particle which we denote by Δx. We cannot say where the particle is to within a wavelength of the wave.

We have $p = \hbar k$. This means that momentum is a wave. In this case, the wave is in what we call "momentum space." As with the

---

[1.] Shakespeare, eat your heart out.

spatial position, we cannot, to within one momentum wavelength, exactly know the momentum. We denote the uncertainty in momentum by Δp. We write the spatial wavefunction as $\psi(x)$ and we write the momentum wavefunction as $\psi(p)$. Fourier transform theory tells us that these two waves are related by:

$$\psi(x) = \frac{1}{\sqrt{2\pi\hbar}} \int_{-\infty}^{\infty} dp\, \psi(p) e^{i\frac{px}{\hbar}}$$

$$\psi(p) = \frac{1}{\sqrt{2\pi\hbar}} \int_{-\infty}^{\infty} dx\, \psi(x) e^{-i\frac{px}{\hbar}} \tag{10.3}$$

Fourier theory also tells us that:

$$\Delta x \Delta p \geq \frac{\hbar}{2} \tag{10.4}$$

This is known as the Heisenberg uncertainty principle[2]. Uncertainty is "caused" by the wave nature of particles. It is impossible to assign a definite spatial position to a wave. Using $p = \dfrac{h}{\lambda}$, $\dfrac{1}{\lambda} = k$, $\hbar = \dfrac{h}{2\pi}$, we have:

$$\Delta x \Delta \frac{h}{\lambda} \geq \frac{h}{4\pi} \Rightarrow \Delta x \Delta k \geq \frac{1}{4\pi} \tag{10.5}$$

It is not the only uncertainty principle in quantum mechanics. Another such uncertainty is:

$$\Delta t \Delta E \geq \frac{\hbar}{2} \tag{10.6}$$

Readers familiar with the momenergy 4-vector of special relativity will correctly recognize these as:

$$\Delta \begin{bmatrix} E \\ p_x \\ p_y \\ p_z \end{bmatrix} \Delta \begin{bmatrix} t \\ x \\ y \\ z \end{bmatrix} \geq \frac{\hbar}{2} \tag{10.7}$$

[2.] W. Heisenberg Z. f. Physik 43, 172, (1927).

Uncertainty of position in space is not just uncertainty of observation; it is an intrinsic property of particles. We are driven to conclude that particles do not follow a definite trajectory through space – that's weird, but, of course, waves do not follow a definite trajectory through space.

In general, there is uncertainty between two operators given by:

$$\Delta a \Delta b \geq \frac{1}{2}\left|\left[\hat{A},\hat{B}\right]\right|. \tag{10.8}$$

Wherein, a is the dynamic variable associated with the operator $\hat{A}$, and b is the dynamic variable associated with the operator $\hat{B}$.

Aside: The reader will in later studies meet the anti-commutator of two operators. This is written as:

$$\{\hat{A},\tilde{B}\}\Psi = \hat{A}\left(\hat{B}(\Psi)\right) + \hat{B}\left(\hat{A}(\Psi)\right) \tag{10.9}$$

EXERCISES

1. What is the commutator of the two operators

   $$\widehat{p_y} = -i\hbar\frac{\partial}{\partial y} \quad \& \quad \hat{x} = x?$$

2. What is the uncertainty relation between the two operators in 1?

3. What is the uncertainty relation between the operators?

   $$\widehat{L_x} = \widehat{y}\widehat{p_z} - \widehat{z}\widehat{p_x} \quad \& \quad \widehat{L_y} = \widehat{z}\widehat{p_x} - \widehat{x}\widehat{p_z} \tag{10.10}$$

4. Calculate $\left[\widehat{p_x},\widehat{x^2}\right]$?

5. Starting with $\Delta t \Delta E \geq \frac{\hbar}{2}$, show that $\Delta t \Delta v \geq \frac{1}{4\pi}$ where v is frequency?

11

ANGULAR MOMENTUM AND THE POSITION OPERATORS

We have said above that the relations between Newtonian dynamic variables are duplicated between the quantum mechanical operators associated with those Newtonian variables. This is the correspondence principle. This allows us to calculate the position operators from the angular momentum operator relations.

The classical expression for angular momentum is:

$$\vec{L} = \vec{r} \times \vec{p}$$
$$L_x = r_y p_z - r_z p_y$$
$$L_y = r_z p_x - r_x p_z \tag{11.1}$$
$$L_z = r_x p_y - r_y p_x$$

The angular momentum operators are thus given by:

$$\widehat{L}_x = \widehat{y}\widehat{p}_z - \widehat{z}\widehat{p}_y$$
$$\widehat{L}_y = \widehat{z}\widehat{p}_x - \widehat{x}\widehat{p}_z \tag{11.2}$$
$$\widehat{L}_z = \widehat{x}\widehat{p}_y - \widehat{y}\widehat{p}_x$$

This leads us to the quantum mechanical operators equivalent of the Newtonian position variables. The position operators are "multiply by the variable" operators. We have:

$$\hat{x} = x$$
$$\hat{y} = y \tag{11.3}$$
$$\hat{z} = z$$

So the x-position operator acts upon a function, $f(x)$, to produce an output function that is the input function multiplied by the variable x:

$$\hat{x}\left(f(x)\right) = x.f(x) \tag{11.4}$$

Aside: Although the linear momentum operators commute with each other, the angular momentum operators do not so commute with each other.

$$\left[\widehat{p_x}, \widehat{p_y}\right] = 0$$
$$\left[\widehat{L_x}, \widehat{L_y}\right] \neq 0 \tag{11.5}$$

11.1 EIGENFUNCTIONS OF THE POSITION OPERATOR (THE DIRAC δ FUNCTION)

The reader might wonder what kind of eigenfunction would be associated with the position operator. The position eigenfunctions are such that:

$$\hat{x}\left(\phi(x)\right) = x.\phi(x) = a_n\phi(x) \tag{11.6}$$

The eigenfunction of the x-position operator is called the Dirac delta function. The Dirac delta function is defined to be an infinitely tall "spike" at the origin on the x-axis which has an integral (area under the spike) of unity. Technically, it is:

$$\delta_{\varepsilon\to 0} = \begin{cases} \dfrac{1}{\varepsilon} & \text{for } -\dfrac{\varepsilon}{2} < x < \dfrac{\varepsilon}{2} \\ 0 & \text{otherwise} \end{cases} \tag{11.7}$$

We have:

$$\int_{-\frac{\varepsilon}{2}}^{\frac{\varepsilon}{2}} dx \; \delta_\varepsilon = \int_{-\frac{\varepsilon}{2}}^{\frac{\varepsilon}{2}} dx \; \frac{1}{\varepsilon} = 1 \qquad (11.8)$$

The delta function is defined as a "spike" at points other than the origin, and it is a function of x, $\delta(x)$. Within an integral, the delta function effectively picks out the value of a function at zero:

$$\int_{-\infty}^{\infty} dx \; \delta(x) f(x) = f(0) \qquad (11.9)$$

By modifying the delta function, we can pick out the value of the function at any point on the x-axis:

$$\int_{-\infty}^{\infty} dx \; \delta(x-a) f(x) = f(a) \qquad (11.10)$$

It is in this way that $\delta(x-a)$ is thought of as representing a particle at position $x = a$. The eigenfunctions of the x-position operator are $\delta(x-a)$ for every (continuous) value of a. The eigenfunctions of the y-position operator and the eigenfunctions of the z-position operator are similarly defined.

The reader might feel that it is a little contrived to construct the Dirac delta function just because the idea of eigenfunction does not make any sense when it comes to position operators. It is the reader's right to so feel if he wishes.

Aside: Paul Adrien Maurice Dirac (1902–1984) made fundamental contributions to both quantum mechanics and quantum electrodynamics including the prediction of anti-matter in his famous Dirac equation[1]. He shared the 1933 Nobel prize with Schrödinger.

As well as being a mathematical genius, Dirac is thought to have had autistic traits. He was well known for the taciturn and precise manner in which he related to people. Einstein once said of him, "…this balancing on the dizzying path between genius and madness…". He

---
[1.] P.A.M.Dirac, Proc Roy. Soc. Lond. A 117, 610 (1928).

was a renowned atheist who, though tolerant of religious people, thought them to be intellectually primitive.

Aside: The simplest representation of the delta-function is:

$$\delta(a) = \frac{1}{2\pi} \int\limits_{-\infty}^{\infty} dx \; e^{iax} \tag{11.11}$$

We have:

$$\phi_g(a) = \frac{1}{2\pi} \int\limits_{-g}^{g} dx \; e^{iax} = \frac{1}{\pi} \frac{\sin(ag)}{a}$$

$$\int\limits_{-\infty}^{\infty} da \; \phi_g(a) = \int\limits_{-\infty}^{\infty} da \; \frac{1}{\pi} \frac{\sin(ag)}{a} = 1 \tag{11.12}$$

But:

$$\phi_g(0) = \frac{g}{\pi}$$

$$\phi_g\left(\pm \frac{\pi}{g} \right) = 0 \tag{11.13}$$

And so, as $g \to \infty$, the contribution to the integral is entirely from $a = 0$.

A NOTE ON ENTANGLEMENT

According to the theory of special relativity, nothing can travel faster than the speed of light. So it is thought that two electrons which are separated in space by, say, a light year, cannot instantly communicate with each other. Indeed, two electrons separated by a light year would require at least a year to communicate with each other. In 1935, Albert Einstein (1879–1955), Boris Podolsky (1896–1966), and Nathan Rosen (1909–1995) presented a paper[1] containing what is now known as the Einstein-Podolsky-Rosen paradox in which they pointed out that, according to quantum mechanics, instantaneous communication is possible between two electrons. Although Einstein, Podolsky, and Rosen considered the position and momentum of two electrons, the situation is more simply described using the spin components of two electrons. We will look at spin more closely later. This was done by David Bohm (1917–1992) in 1951[2].

Bohm postulated that a system with zero spin decays into two spin $\frac{1}{2}$ particles that fly apart from each other. When the particles are distantly separated, an observer measures the z-component of spin of particle A to be up. This means that (conservation of angular momentum) we know that the z-component of spin of particle B is

---

1. A.Einstein, B.Podolski, and N.Rosen Phys. Rev. 47 777 (1935).
2. D. Bohm: Quantum Theory Pub: Prentice Hall (1951).

down. Suppose that each particle has its own wavefunction. Measuring the z-spin of one particle has caused the other particle to collapse its wavefunction and to collapse in a specific, non-probabilistic way. Wavefunctions do not collapse non-probabilistically in quantum mechanics – we must have made a mistake. Let us suppose that, instead of two wavefunctions, one for each particle, there is only one wavefunction governing both particles. In this scenario, when we measure the z-spin of one particle, the single wavefunction of the two particles collapses into a particular eigenstate. Yet, still, there is instant communication from particle A to particle B saying into which eigenstate the particles have collapsed.

In classical mechanics, each particle has a definite angular momentum from the very start. Measuring the angular momentum of one particle does not affect the other particle. It is because, in quantum mechanics, systems exist in a superposition of eigenstates that we need instant communication when the spatially extensive wavefunction collapses. This phenomenon of "instant communication" is called entanglement or, sometimes, quantum entanglement.

Entanglement is not concerned with only two separate particles. When a photon of light hits a 45° polarizer, it becomes a wavefunction that is a superposition of having passed through the polarizer and having been reflected from the polarizer. If, after, say, a year, the photon is detected to have passed through the polarizer, then the wavefunction collapse is communicated instantly to the other side of the polarizer.

We go back to the two particles. Our two particles are now in definite states with regard to the z-component of spin. Suppose we now measure the x-component of spin of the A particle. When we do this, because the spin operators $\widehat{S_z}$ and $\widehat{S_x}$ are not commutative, we destroy knowledge of the z-component of spin eigenstate. Thus, particle B, which was definitely in the down z-spin state is no longer in that definite state. The state of the distant particle has been instantly affected by a local measurement of x-spin. Instant communication, which entanglement implies, is an irremovable feature of quantum mechanics, and it will remain so while superposition remains as a feature of quantum mechanics.

Of course, if one adopts the view that a superposition of states is outside of time (there is no time in the complex plane), then instant communication is to be expected. Special relativity is a theory based in the hyperbolic complex numbers, $\mathbb{S}$, and does not apply to the Euclidean complex numbers, $\mathbb{C}$.

12.1   THE BELL INEQUALITIES

It might occur to the reader that perhaps the spin components of the two particles introduced above are actually set when the particles first separate as some function of one or more variables that are hidden from our observation. This circumvents the idea of wavefunction collapse instantly communicating from one place to another. Such theories are called hidden variable theories. It was shown in 1964 by J. S. Bell[3] that such theories need to satisfy particular mathematical constraints called "Bell's inequalities." Experiments by Clauser and Shimony[4] in 1978 and by Alain Aspect et al in 1981[5] and 1982[6] have now ruled out the possibility of any local hidden variables.

[3] J.S.Bell: Physics 1 195 (1964).

[4] J.F.Clauser & A.Shimony Bell's Theorem: experimental tests and implications. Reports on progress in Physics 41, 1881, (1978).

[5] Aspect et al. Experimental tests of realistic local theories via Bell's theorem. Phy Rev Lett 47, 460, (1981).

[6] A. Aspect, J. Dalibard, & C. Roger. Physical Review Letters Vol 49 pgs 91 & 1804 : December 1982.

13

EXPECTATION VALUE AND STANDARD DEVIATION

When we measure a property, say the angular momentum, of a physical system, say an electron in orbit around an atomic nucleus, we get an eigenvalue of the appropriate operator. If we immediately measure the same system again, we will get the same eigenvalue because the first measurement "collapsed" the system into the particular eigenstate associated with that eigenvalue[1]. Immediately after a measurement is made, a physical system is not in a superposition of all possible states but is in the eigenstate that was just observed. If we thus successively and immediately measure the system a hundred times, we will get the same eigenvalue a hundred times.

If we measure a hundred similar physical systems, say a hundred hydrogen atoms - a hundred electrons in orbit around a hundred protons – we will get a hundred eigenvalues, but they might not all be the same. If we sum these hundred eigenvalues and divide by a hundred, we will have the average value of the dynamic variable associated with the eigenvalues. In quantum mechanics, this average value of a variable within a large number of similar physical systems is called the expectation value of that variable. The reader

---

[1] If we wait a while before we re-measure the system, the system will have evolved into a superposition again.

should note that the expectation value (average value) is unlikely to be a particular eigenvalue of any of the many systems that were measured.

The expectation value, $\langle \hat{A} \rangle$, of the dynamic variable associated with the operator $\hat{A}$ within a physical system described by the wavefunction ψ is given by:

$$\langle \hat{A} \rangle = \int_{-\infty}^{\infty} dx \ \psi^{\circ} \hat{A}(\psi) \tag{13.1}$$

For example, the expectation value of x-momentum is given by:

$$\langle \hat{p_x} \rangle = -i\hbar \int_{-\infty}^{\infty} dx \ \Psi^{*}(t,x) \frac{\partial \Psi(t,x)}{\partial x} \tag{13.2}$$

And, the expectation value of x-position is given by:

$$\langle \hat{x} \rangle = \int_{-\infty}^{\infty} dx \ \Psi^{\circ}(t,x)x.\Psi(t,x)$$

$$= \int_{-\infty}^{\infty} dx \ x.|\Psi|^2 \tag{13.3}$$

Aside: We note that if the expectation value of the position of a particle is changing, then the particle is moving. In other words, a particle is moving if:

$$\frac{\partial \langle \hat{x} \rangle}{\partial t} \neq 0 \tag{13.4}$$

If Ψ is a function of multiple spatial variables, then the expectation value is a multiple integral over every spatial variable.

As another example, the expectation value of energy of a bound state (stationary state) is given by:

$$\langle \hat{E} \rangle_n = \int_{-\infty}^{\infty} dx \ \phi_n^{\circ} \hat{H} \phi_n = E_n \int_{-\infty}^{\infty} dx \ \phi_n^{\circ} \phi_n \tag{13.5}$$

The energy expectation for a wave packet:

$$\Psi(t,x) = \sum_{n=1}^{n=\infty} a_n \phi_n e^{-i\frac{E_n}{\hbar}t} \tag{13.6}$$

is given by:

$$\langle \hat{E} \rangle = \int_{-\infty}^{\infty} dx \ \Psi(t,x)^* \hat{H} \Psi(t,x) = \sum_{n=1}^{n=\infty} |a_n|^2 E_n \qquad (13.7)$$

Notice that this energy expectation value is constant in time – energy is conserved.

WORKED EXAMPLE

What is the energy expectation value of the wavefunction $\Psi = (a+ib)\psi_1 + (c+id)\psi_2$?

We have:

$$\Psi = (a+ib)\psi_1 + (c+id)\psi_2$$
$$\Psi^* = (a-ib)\psi_1^* + (c-id)\psi_2^* \qquad (13.8)$$
$$\hat{H}\Psi = (a+ib)E_1\psi_1 + (c+id)E_2\psi_2$$

The expectation value is given by:

$$\langle \hat{H} \rangle = \int_{x=0}^{x=\infty} dx \Psi^* \hat{H}\Psi \equiv \sum_{n=1}^{n=2} \Psi^* \hat{H}\Psi$$

$$= \left((a-ib)\psi_1^* + (c-id)\psi_2^* \right)\left((a+ib)E_1\psi_1 + (c+id)E_2\psi_2 \right)$$

$$= (a-ib)\psi_1^*(a+ib)E_1\psi_1 + (a-ib)\psi_1^*(c+id)E_2\psi_2 \qquad (13.9)$$

$$+(c-id)\psi_2^*(a+ib)E_1\psi_1 + (c-id)\psi_2^*(c+id)E_2\psi_2$$

$$= (a-ib)(a+ib)E_1.1 + (a-ib)(c+id).0$$

$$+(c-id)(a+ib).0 + (c-id)(c+id)E_2.1$$

$$= |a+ib|^2 E_1 + |c+id|^2 E_2$$

13.1 STANDARD DEVIATION AND UNCERTAINTY RELATIONS

We can form the product (square) of an operator, $\hat{A}$, with itself to produce another operator, $\widehat{A^2} = \hat{A}(\hat{A})$. This operator will have an expectation value associated with it, $\left\langle \widehat{A^2} \right\rangle$. We can simply square the expectation value of the operator $\widehat{A}$ to get $\left\langle \widehat{A} \right\rangle^2$. Statisticians can put these two calculations together to get the standard deviation associated with the operator $\widehat{A}$. That standard deviation is:

$$\Delta \widehat{A} = \sqrt{\left\langle \widehat{A^2} \right\rangle - \left\langle \widehat{A} \right\rangle^2} \tag{13.10}$$

This is a measure of the spread of the eigenvalues of the operator $\widehat{A}$. The quantum mechanical equivalent of the standard deviation is the uncertainty denoted $\Delta \widehat{A}$.

WORKED EXAMPLE

What is the uncertainty in the energy of a system in an energy eigenstate?

$$\Delta \widehat{E_n} = \sqrt{\left\langle \widehat{E_n^2} \right\rangle - \left\langle \widehat{E_n} \right\rangle^2} \tag{13.11}$$
$$= \sqrt{E_n^2 - E_n^2} = 0$$

We see that there is no uncertainty in the energy of a physical system that is in an energy eigenstate – we knew that actually.

When we calculate the expectation values for position and for momentum, we get:

$$\Delta x = \frac{a}{\sqrt{2}} \quad \& \quad \Delta p_x = \frac{\hbar}{a\sqrt{2}} \tag{13.12}$$

This leads to the famous Heisenberg's uncertainty relation:

$$\Delta x \Delta p_x = \frac{a}{\sqrt{2}} \frac{\hbar}{a\sqrt{2}} = \frac{\hbar}{2} \tag{13.13}$$

WORKED EXAMPLE

1. What is the uncertainty in energy of wavefunction $\Psi = (a + ib)\psi_1 + (c + id)\psi_2$?

We have from above:

$$\langle \widehat{H} \rangle = |a + ib|^2 E_1 + |c + id|^2 E_2 \qquad (13.14)$$

Now:

$$\langle \widehat{H^2} \rangle = \sum_{n=1}^{n=2} \Psi^* \widehat{H^2} \Psi \qquad (13.15)$$

$$= \left((a - ib)\psi_1^* + (c - id)\psi_2^* \right)$$

$$\left((a + ib)(E_1)^2 \psi_1 + (c + id)(E_2)^2 \psi_2 \right)$$

$$= |a + ib|^2 (E_1)^2 + |c + id|^2 (E_2)^2$$

Giving:

$$\Delta H = \left[\langle \widehat{H^2} \rangle - \langle \widehat{H} \rangle^2 \right]^{\frac{1}{2}}$$

$$= \left[|a + ib|^2 E_1^2 + |c + id|^2 E_2^2 - \left(|a + ib|^2 E_1 + |c + id|^2 E_2 \right)^2 \right]^{\frac{1}{2}} \qquad (13.16)$$

13.2 EHRENFEST'S THEOREM

Ehrenfest's theorem deals with the time evolution of expectation values. We remind the reader that the expectation value of a variable associated with an operator, $\widehat{A}$ is given by the overlap integral (inner product):

$$\langle \widehat{A} \rangle = \int dx \ \Psi^* \widehat{A}(\Psi) \qquad (13.17)$$

Most expectation values do not change through time, but some do. We calculate how an expectation value might evolve with time by taking the derivative with respect to time. For an operator, $\widehat{A}$, we have:

$$\frac{\partial}{\partial t} (\langle \widehat{A} \rangle) = \frac{\partial}{\partial t} \left(\int dx \ \Psi^* \widehat{A}(\Psi) \right) \qquad (13.18)$$

We can differentiate under the integral sign to get:

$$\frac{\partial}{\partial t} \langle \widehat{A} \rangle = \int dx \left(\frac{\partial \Psi^*}{\partial t} \widehat{A}(\Psi) + \Psi^* \frac{\partial \widehat{A}}{\partial t} \Psi + \Psi^* \widehat{A} \frac{\partial \Psi}{\partial t} \right) \qquad (13.19)$$

Using the time dependent Schrödinger equation, TDSE, we get:

$$\widehat{H}(\Psi) = i\hbar \frac{\partial \Psi}{\partial t} \quad : \quad \widehat{H}(\Psi^*) = i\hbar \frac{\partial \Psi^*}{\partial t}$$

$$\frac{\partial}{\partial t}\langle \widehat{A}\rangle = \frac{1}{i\hbar}\int dx \left[-\widehat{H}(\Psi^*)\widehat{A}(\Psi) + \Psi^*\widehat{A}(\widehat{H}(\Psi)) \right] + \left\langle \frac{\partial \widehat{A}}{\partial t}\right\rangle \quad (13.20)$$

We can rearrange (using the hermicity of $\widehat{H}$):

$$\frac{\partial}{\partial t}\langle \widehat{A}\rangle = \frac{1}{i\hbar}\int dx \left(\Psi^*\left[\widehat{A},\widehat{H}\right](\Psi)\right) + \left\langle \frac{\partial \widehat{A}}{\partial t}\right\rangle \quad (13.21)$$

This last statement is known as Ehrenfest's theorem.

13.3 CONSERVATION LAWS

For operators that are not time-dependent, the right-most term of Ehrenfest's theorem is zero. If the operator $\widehat{A}$ commutes with the Hamiltonian, $\widehat{H}$, the whole expression is zero and we have:

$$\frac{\partial}{\partial t}\langle \widehat{A}\rangle = 0 \quad (13.22)$$

The expectation value of any operator is a constant if the operator commutes with the Hamiltonian. Most of the operators in quantum mechanics commute with the Hamiltonian.

We see that $\langle \widehat{A}\rangle$ is a constant of the system – it is a conserved quantity – we have a conservation law.

So, there you are; if an operator commutes with the energy operator (Hamiltonian), the Newtonian dynamic variable with which that operator is associated is a conserved quantity.

13.4 THE HEISENBERG AND SCHRÖDINGER REPRESENTATIONS

There are two ways of doing quantum mechanics. One of these ways is called wave mechanics or the Schrödinger representation; the other of these ways is called matrix mechanics or the

Heisenberg representation. Within the Schrödinger representation, it is the state, $\psi(t, x)$, that evolves in time, and we have:

$$\left\langle \psi^{\circ}(t) \middle| \widehat{A} \middle| \psi(t) \right\rangle = \left\langle \psi^{\circ}(0) \middle| U^{\dagger} \widehat{A} \ U \middle| \psi(0) \right\rangle \tag{13.23}$$

The Schrödinger equation has the term $\dfrac{\partial \psi(t,x)}{\partial t}$ relating to the time evolution of the state, $\psi(t, x)$, and we see that the state, ψ, is a function of time.

Within the Heisenberg representation, it is the operator that evolves in time, and we have:

$$\widehat{A(t)} = U^{\dagger} \ \widehat{A(0)} \ U \tag{13.24}$$

Corresponding to the Schrödinger equation of motion of the Schrödinger representation, there is a Heisenberg equation of motion:

$$\frac{d\widehat{A(t)}}{dt} = i \left[\widehat{H}, \widehat{A(t)} \right] \tag{13.25}$$

Wherein $\widehat{A(t)}$ is an operator that is dependent upon time and $\widehat{H}$ is the Hamiltonian.

We can have different representations because we observe inner products and not operators or wavefunctions.

The Schrödinger equation defines the state of a physical system in terms of the spatial positions of the parts of the system. There are some things in physics, and intrinsic spin is one of these things, that cannot be described by spatial positions. This is why the Schrödinger equation is of no use when we deal with intrinsic spin.

13.5 BRAS AND KETS (ADDENDUM)

The expectation value written in Dirac notation is:

$$\left\langle \widehat{A} \right\rangle = \left\langle \Psi \middle| \widehat{A} \middle| \Psi \right\rangle \tag{13.26}$$

EXERCISES

1. A particle has wavefunction $\Psi(x) = \sqrt{\dfrac{2}{a}} \sin\left(\dfrac{\pi}{a} x\right)$.. Find the expectation value of the momentum operator $\widehat{p_x} = -i\hbar \dfrac{\partial}{\partial x}$ and the position operator $\hat{x} = x$.

2. If $\psi(x) = \dfrac{1}{x^4}$, what is the expectation value of the position operator $\langle \hat{x} \rangle$? You will need to know that the integral of odd functions is zero. What is the expectation value of the operator $\langle \widehat{x^3} \rangle$? Is $\psi(x) = \dfrac{1}{x^4}$ square integrable? If not, is it a valid wavefunction?

14

PROBABILITY

The quantum mechanical view is that, when a physical system is unobserved, it is in a superposition (linear sum) of possible states and that, when the physical system is observed, it collapses into a single one of the possible states. The reader might think, as any reasonable person might think, that somewhere within the superposition of possible states, there is a little something, a variable, that determines into which single one of the possible states the physical system will collapse and that, if we knew what this variable was, we could predict which of the possible states would be "the chosen one." There is no such hidden variable, or, at least, there is no such hidden variable that is local to the physical system. This was shown by J. S. Bell in 1964[1].

Which of the many possible states a physical system will collapse into when it is observed is, as far as we know, determined purely by chance. It seems that "God does play dice." Let us consider the hydrogen atom, and let us consider the energy of an electron in orbit in that atom. As far as the energy is concerned, the unobserved electron in the hydrogen atom is described by a set of weighted basis eigenfunctions of the hydrogen energy operator linearly added together (to form a single complete energy wavefunction) – a linear sum. We see this written as:

$$\Psi = c_1\psi_1 + c_2\psi_2 + c_3\psi_3 + ... \qquad (14.1)$$

---

[1]. J.S.Bell On the Einstein, Podolsky, Rosen Paradox. Physics, 1, 3, 195–200 (1964).

Where $c_n \in \mathbb{C}$ are just complex constants that determine the complex amplitude of the corresponding eigenfunction of the Hamiltonian in the "complete" wavefunction, Ψ. These eigenfunctions are orthogonal to each other under the overlap integral inner product. We can, and do, scale down the coefficients, c_i, to normalize this wavefunction so that the sum of the moduli of these coefficients is unity:

$$\sum |c_i|^2 = 1 \qquad (14.2)$$

When it is observed, the electron collapses into a single one of these eigenfunctions and the electron is described by the eigenfunction:

$$\Psi = \psi_n \qquad (14.3)$$

If we take the inner product (overlap integral) of the complete wavefunction, (14.1), and the collapsed wavefunction,(14.3), we get:

$$\int dx \ \Phi^* \Psi =$$
$$\int dx \ (c_n \psi_n)^* c_1 \psi_1 + \int dx \ (c_n \psi_n)^* c_2 \psi_2 + ... + \int dx \ (c_n \psi_n)^* c_n \psi_n + ... \qquad (14.4)$$

Because the basis eigenfunctions are orthogonal, this becomes:

$$\int dx \ \Phi^* \Psi = 0 + 0 + ... + \int dx \ (c_n \psi_n)^* c_n \psi_n + ... + 0$$
$$= c_n^* c_n \int dx \ \psi_n^* \psi_n \qquad (14.5)$$
$$= c_n^* c_n$$

The modulus squared of the coefficient is proportional to the probability that, if we observed the electron, we would find it in the state ψ_n (with energy E_n).

$$P_{\psi_n} \propto |c_n|^2 = c_n^* c_n \qquad (14.6)$$

In fact, because, in anticipation of the involvement of probability, we normalized the complete wavefunction, the proportionality constant is unity, and we have.

$$P_{\psi_n} = |c_n|^2 = c_n^* c_n \qquad (14.7)$$

A complete wavefunction is the weighted linear sum of different (complex) amounts of each basis eigenfunction. The "size" of the amount (the modulus of the complex coefficient) determines the probability that a particular eigenfunction will be "the chosen one" when the electron described by the complete wavefunction is observed and collapses into a possible state described by one eigenfunction.

Let us pull the above together; if a physical state is described by the complete wavefunction, Ψ, the probability that it will be observed to be in the physical state described by the basis eigenfunction, Φ_n, is the normalized square of the overlap integral:

$$P = \frac{1}{A^2}\langle \Phi_n | \Psi \rangle^2 = \frac{1}{A^2}\left|\int_{-\infty}^{\infty} dx\, \Phi_n{}^*\Psi\right|^2 \qquad (14.8)$$

A is just a constant that normalizes the complete wavefunction. If we know the possible outcomes of an experiment, then we know the basis eigenfunctions of the wavefunction (we know the solutions of the TISE, perhaps). There is one basis eigenfunction for each possible outcome. The above formula allows us to predict the likelihood of each possible outcome. If we do the experiment a million times, we can predict how many of the million outcomes will give a particular result.

The "understanding" that the modulus of the squared wavefunction $|\Psi|^2$ corresponds to probability density was first voiced by Max Born (1882–1970) and is known as the Born interpretation[23].

Aside: Max Born was a German/British Jew. He spent most of his early academic career at Gottingen alongside noteworthy individuals such as Hilbert, Klien, and Minkowski until, when the Nazis came to power in Germany in 1933, he was dismissed from his post and moved to Cambridge, England. It was Born who taught matrix algebra to Heisenberg thereby contributing to the matrix formulation of quantum mechanics, and to was to Born that Einstein wrote the famous letter asserting the "God does not play dice."

Born won the 1954 Nobel prize for physics for his contributions to quantum theory.

[2.] M Born Z. f. Physik 37, 863, (1926) and M Born Z. f. Physik 38, 803, (1926).
[3.] M Born Nature 119 354 (1927).

14.1   THE PROPERTIES OF PROBABILITY

In quantum mechanics, probability is treated in a way similar to the way we treat electrical charge. We have concepts like a probability current, a probability continuity equation, conservation of probability, and a probability density. It stretches the brain to wonder why probability can be treated as if it were a kind of charge. Perhaps nature is telling us something that we seem unable to hear clearly. With this "probability charge" concept in mind, we proceed.

For a given wavefunction, $\psi(t, x, y, z)$, the probability density is defined as:

$$\rho(x,t) = \psi^*(x,t)\psi(x,t) = |\psi|^2 \tag{14.9}$$

Different values of the spatial and time co-ordinates give an amount of probability at each point in space at each point in time. At any particular point in time, probability density is a scalar field.

The probability current is defined as:

$$j = -i\frac{\hbar}{2m}\left(\psi^*\frac{\partial^2\psi}{\partial x^2} - \psi\frac{\partial^2\psi^*}{\partial x^2}\right) \tag{14.10}$$

By differentiating (14.9) with respect to time and with respect to space, and remembering that both $\{\psi, \psi^*\}$ are solutions of the TDSE, we are led to the probability continuity equation:

$$\frac{\partial\rho}{\partial t} + \frac{\partial j}{\partial x} = 0$$
$$\frac{\partial\rho}{\partial t} + \nabla\bullet j = 0 \tag{14.11}$$

We have given both the 1-dimensional and the 3-dimensional versions. The probability continuity equation expresses the conservation of probability. Total probability is always unity, and so probability must be conserved.

14.2   PROBABILITY DENSITY AND PARTICLE POSITION

If $\Psi(t, x)$ is an acceptable solution of the TDSE for a particle, then the probability density multiplied by Δx, $|\Psi|^2 \Delta x = \Psi^*\Psi\Delta x$, is the probability of the particle being observed to be in the interval $[x, x + \Delta x]$. We have:

$$P_{\text{particle being in } dx=[x,x+\delta x]} = |\Psi(t,x)|^2 \Delta x \qquad (14.12)$$

The multiplication is most often done through integration.

$$P_{[a,b]} = \int_{x=a}^{x=b} dx \, |\Psi|^2 \qquad (14.13)$$

Since the particle must be somewhere between plus infinity and minus infinity, we have:

$$\int_{x=-\infty}^{x=\infty} dx \, |\Psi|^2 = 1 \qquad (14.14)$$

WORKED EXAMPLE

A probability density is given by $\rho(x) = \dfrac{1}{x^4}$. What is the probability that the particle will be found in the interval [2, 3]?

We have:

$$P = \int_{2}^{3} dx \, \frac{1}{x^4} = \left[-\frac{1}{3x^3} \right]_{2}^{3} = \frac{19}{648} \qquad (14.15)$$

EXERCISES

1. A probability density is given by $\rho(x) = \dfrac{2}{x^3} + \dfrac{x}{5}$. What is the probability that the particle will be found in the interval [2, 3]

2. For the wavefunction
$$\psi(x) = \sqrt{\frac{1}{a}} \sin\left(\frac{5\pi}{a}x\right) + \frac{i}{2}\sqrt{\frac{1}{2a}} \sin\left(\frac{3\pi}{a}x\right) - \sqrt{\frac{2}{a}} \sin\left(\frac{4\pi}{a}x\right),$$ what is the probability that, when measured, we will observe the physical system to be in a state corresponding to $\sin\left(\frac{5\pi}{a}x\right)$?
Hint: is $\psi(x)$ normalized.

CHAPTER 15

THE HISTORY OF QUANTUM MECHANICS

If, given the heavy going of the previous few chapters, the reader is no longer sitting comfortably, then the reader needs a somewhat easier chapter to relax her into the comfort in which she started this book. We present that easier chapter now.

As with most things in life, quantum mechanics did not spring overnight into being a full and glorious part of humanity's world understanding. It was dragged and shoved into being by the failures of 19th century physics and by the effulgent insights of a few individuals and by a lot of doubt and a lot of hard graft. It developed over many years, and it is still developing today. In this chapter, we briefly summarize some of the pangs of its birth.

15.1 NEWTONIAN MECHANICS

Newtonian mechanics is about the behavior of electrically neutral particles of matter. Sometimes the particles of Newtonian mechanics are large particles like planets orbiting the sun. Other times, the particles are the size of billiard balls, but all non-microscopic electrically neutral particles behave as described by Newtonian mechanics. By the middle of the 19th century, Newtonian mechanics was firmly established as the set of rules that govern the mechanics of particles.

This was so even though the Newtonian mechanics of the solar system was known to be not perfect. In 1859, Urbain Le Verrier (1811–1877) reported a discrepancy between the predicted orbit of the planet Mercury and the observed orbit, but this was thought to be caused by the presence of a yet undiscovered planet closer to the Sun than Mercury. Such was the confidence in Newtonian mechanics that the unseen planet was named Vulcan after the Roman god of fire (a black-smith) in anticipation of it being discovered.

Newtonian mechanics has served, and continues to serve, humankind very well. It is used to plant space-probes gently on to the surface of distant worlds and to engineer marvellous bridges across wide waterways. However, in the early 20th century it became subject to a "double whammy" in that it was shown to be both insufficient to deal with objects moving at high velocities and insufficient to deal with very small objects. It was replaced by relativistic mechanics in the case of the rapidly moving objects, and it was replaced by quantum mechanics in the case of the very small objects.

Aside: Newtonian mechanics was subject to one more "whammy" in 1915 when it was replaced by the mechanics of general relativity.

15.2 ENERGY AND MOMENTUM AND WAVES

Newtonian mechanics is formulated in terms of energy and momentum. In Newtonian mechanics, force is temporal rate of change of momentum, or spatial rate of change of energy, and the motion of a particle is defined by its energy and its momentum through the conservation of energy and the conservation of momentum. (In relativistic physics these are combined into one entity called momenergy.) The concepts of momentum and energy are also central to the theory of quantum mechanics. They are also central to relativity theory and to quantum field theory and to general relativity. These well-aged Newtonian concepts are still with us even in the most advanced of our physical theories. There is a good reason for this; momentum is connected, through its conservation, to the homogeneity of empty space. That homogeneity is the invariance under translation in space of a physical system. Since physics is invariant under spatial translation for all our physical theories, we will have momentum in all our physical theories. Energy is just momentum in the time direction, and, since time is homogeneous, we will have energy in all our physical theories.

Within Newtonian mechanics, the motion of waves is described by wavelength or frequency and by the propagation vector. These concepts are not energy and momentum, and 19th century physicists did not view waves as having energy and momentum in the same way that particles have energy and momentum. So it was to 19th century physicists that there were two separate types of things in the universe; these two types of things were particles with energy and momentum and waves with frequency and a propagation vector.

15.3 THE NATURE OF LIGHT

In his day, Newton (1642–1727) had assumed light to be corpuscular (particles), but, by the middle of the 19th century, Newton's view of the corpuscular nature of light had been superseded by the understanding that light, which is just a type of electromagnetic radiation, is a wave. It would be wrong to say that Newton's corpuscular view was the generally accepted view even in his own time. Many of Newton's contemporaries including Robert Hooke (1635–1703), Christaan Huygens (1629–1695), and Leonhard Euler (1707–1783) held the view that light was a wave. Others that followed, including Pierre-Simon Laplace (1749–1827), agreed with Newton's corpuscular view.

In 1803, Thomas Young (1773–1829) conducted experiments demonstrating interference patterns generated by light passing through two very small holes very close to each other[1]. This experiment is known as Young's double slit experiment because it is normally repeated with slits rather than with holes.

---

[1] Reported in "On the nature of light and colors" by Thomas Young. 1803.

15.4 DOUBLE SLIT EXPERIMENT

If I fire bullets at a wall with two slits in it, most bullets will bounce back from the wall but some will go through one or other of the two slits. Each bullet will go through only one slit. If I place a fence behind the wall to mark where the bullets passed through the wall, I will get two separate lines of bullet holes in the fence corresponding to the two slits. Such behavior is the behavior of particles.

If I now send a wave, say a water wave, against the two slits, most of the wave will be reflected from the wall but some part of the wave will go through each of the two slits. Having passed through the slits, the bits of wave will each start to spread out from their slit in all directions. Eventually, the two bits of wave will meet each other and will interfere with each other. When the interfering waves hit the fence, the positions of their crests will give a set of several (more than two) lines. Such behavior is the behavior of waves.

If I now send lots of electrons against the two slits in the wall, I get a set of several lines on the fence indicating that the many electrons are a wave. If I send a single electron, it will appear in only one place on the fence, but where it appears will correspond to a place in one of the several interference lines indicating that it is a single particle that has interfered with itself and thus has wave-like properties. Electrons are thus both particles and waves!

Since interference is a wave phenomenon, with his double slit experiment, Young proved that light was a wave, but this was not accepted until 1817 when Augustin-Jean Fresnel (1788–1827) won the Academie des Sciences prize for his thesis that light was a wave. Even then, the wave-like nature of light would not have been accepted if not for the experiment of Francois Arago (1786–1853) in which he observed the Arago spot predicted by Simeon Poisson (1781–1840) from Fresnel's thesis. Ironically, Poisson, who was confident that the Arago spot would not be observed, had made the prediction as a means of showing Fresnel's wave thesis to be ridiculous.

In 1861 and 1862, James Clerk Maxwell (1831–1879) produced the Maxwell equations of electromagnetism. From these, the wave equations of electromagnetic radiation can be deduced, and, also from these, it is clear that light is such an electromagnetic wave. If one accepts the Maxwell equations, and everyone did and still does, then light is a wave specified by a frequency and a propagation vector. In 1887, Heinrich Hertz (1857–1894) detected the electromagnetic waves predicted by the Maxwell equations. There we have it; how can you dispute that? The reader will see that to have proposed in those times that light has a particle like nature would have seemed insane.

There was a problem with the wave nature of light. How did light pass through empty space from the sun to the Earth? Waves need a medium in which to undulate. The proposed solution was the aether, but the story of that is straying too far from the theme of this book[2].

By the latter part of the 19th century, the Newtonian mechanics of particles and the wave nature of light were happily co-existing. Both seemed to have been verified by experiment and were well understood. It was simple; matter is particles specified by energy and momentum; electromagnetic radiation is waves specified by a frequency and a propagation vector. This understanding is called classical physics.

Things were not that cosy for long. It gradually became apparent to physicists of the time that there were physical phenomena that could not be explained by the scenario that matter is particles and electromagnetic radiation is waves. These phenomena included blackbody radiation, the photo-electric effect, Compton scattering, and eventually the nature of the Rutherford atom. As early as 1885, Johann Jakob Balmer (1825–1898) had shown that electromagnetic radiation emitted by hydrogen atoms came in discreet frequencies which he had described by the formula:

$$\omega = N\left(\frac{1}{2^2} - \frac{1}{n^2}\right) \quad : \quad n = 3, 4, 5, \ldots \tag{15.1}$$

The Balmer series of spectral lines of the hydrogen atom is named after him.

[2.] See "Empty space is amazing stuff" by Dennis Morris ISBN: 978-0-9549780-7-5

15.5 BLACK BODY RADIATION

Technically, a black body is a body of matter that is in thermal equilibrium with electromagnetic radiation at a given temperature. A black-body is an idealized perfect emitter of electromagnetic radiation and a perfect absorber of electromagnetic radiation. Perfect examples do not exist in nature, but they do exist in theoretical physics.

So, a black-body emits light at all frequencies less than a high frequency cut-off while reflecting none of the light that falls upon it. So why does a hot poker appear red or white and a cold poker appear black? It is the amount of light that is emitted at each frequency that determines the color of the poker. When a poker is glowing white hot, it is emitting a lot of light with wavelengths in the visible spectrum (circa $390 \times 10^{-9}\ M - 700 \times 10^{-9}\ M$). When a poker is cold and black, it is emitting very little light at these wavelengths. In that state, it is emitting most of its light at wavelengths outside of, and of less frequency than, the visible spectrum. If we draw a graph of the amount of light being emitted (the energy density to be precise) at each wave-length against the wave-length of the emitted light, we get something like the graph below.

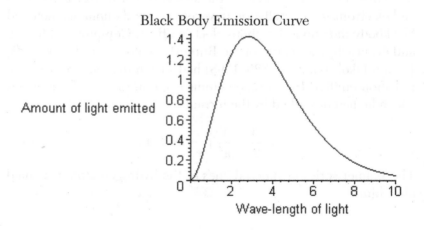

Black Body Emission Curve

We see that there is a peak of emission at a particular wave-length. At different temperatures, the peak appears at different wave-lengths. For a white hot poker, this peak wave-length is the wave-length that corresponds to white light. For a red hot poker, this peak wave-length is the wave-length that corresponds to red light. For a cold black poker, this peak wave-length is outside of the visible spectrum. Plots like the one given above were found experimentally. The problem is to explain these experimental results theoretically. This is a problem because the blackbody spectrum cannot be explained using classical physics.

According to classical physics (matter is particles and electromagnetic radiation is waves), the exchange of electromagnetic radiation and matter should be continuous, and so we should get no high frequency cut-off. Indeed, according to classical physics, we should get most of the electromagnetic radiation emitted at very high frequencies. This is not what is observed. Classical physics was unable to explain the experimentally observed spectrum of black bodies. The best that could be done was the Rayleigh-Jeans formula, presented in 1905 by Lord Rayleigh[3] (1842–1919) and Sir James Jeans (1877–1946), that gives the energy density, $u(v, T)$, as a function of frequency and temperature as:

$$u = \frac{8\pi v^2}{c^3} KT \qquad (15.2)$$

Where v is the frequency of the emitted electromagnetic radiation, K is Boltzmann's constant, T is the temperature in degrees absolute, and c is the speed of light.

This formula was derived from classical physics. This formula does not fit the experimental data; indeed, if one integrates over all frequencies to get the total energy density, one gets infinity. Therefore, classical physics is wrong.

The Rayleigh-Jeans formula works reasonably well at low frequencies, but it is clearly in error at high frequencies. The error of the Rayleigh-Jeans formula was known to classical physics as the *ultraviolet catastrophe*[4]. So, how do we explain the black-body experimental data?

[3.] Lord Rayleigh's given name was John William Strutt.

[4.] With titles like this, it is no wonder that someone invented science fiction.

The explanation was presented by Max Planck[5] (1858–1947) in 1900[6] and 1901[7], four years before the Rayleigh-Jeans formula was presented. Planck assumed that, at a specific frequency, ν, energy comes in discreet amounts called quanta, and that the amount of the energy in a quantum of energy at a particular frequency is given by:

$$E = \nu h = \omega \frac{h}{2\pi} \qquad (15.3)$$

Here, $\omega = 2\pi\nu$, and h is a constant of nature known today as Planck's constant. Planck's constant is:

$$h = 6.62 \times 10^{-34} \text{ Joule Seconds} \qquad (15.4)$$

The reader should be aware that we often use $\hbar$ instead of h where:

$$\hbar = \frac{h}{2\pi} \qquad (15.5)$$

Aside: Max Karl Ernst Ludwig Planck was awarded the Nobel prize for physics in 1918.

Aside: Theoretical physics uses mass-dimensions rather than Joules or seconds. Based on the theory of special relativity, we take the mass-dimension of time to be the same as the mass-dimension of space which we write as:

$$[L] = [T] \qquad (15.6)$$

Where L means length and T means time. We are measuring time in the same units that we use to measure length. We might say that the Earth is 8 seconds from the sun. Based on this, we see that the dimensions of velocity (meters per second) is length divided by time which is just a number, and so the physical constant that is the velocity of light is just a number. With this in mind, based on $E = mc^2$, we take the mass-dimension of energy to be the same as that of mass:

$$[E] = [M] \qquad (15.7)$$

[5.] M Planck Ann Phys 1,69, (1900).
[6.] M Planck, Verh Dtsch Phys. Ges 2, 244 (1900).
[7.] M. Planck, Ann. Physik, 4, 553 (1901).

Thus Planck's constant has mass-dimensions:

$$[h] = [M][T] \qquad (15.8)$$

Since Planck's constant is a constant of nature, like the velocity of light, we feel that it ought to be no more than a number, and so we take it that the mass-dimensions of mass are the inverse of the mass-dimensions of time. This is:

$$[L] = [T] = [M]{-}1 = [E]{-}1 \qquad (15.9)$$

Which makes Planck's constant no more than a number.

Note: In other cases, we have to include the mass-dimension of electric charge as a separate mass-dimension.

Planck's view is that radiation comes in streams of particles, called quanta. Note that Planck's formula (15.3) does not say energy is quantitised. In fact, it speaks of only electromagnetic energy, and it does not say electromagnetic energy is quantitised. It says that, at a specific frequency, electromagnetic energy comes in quanta. At a very slightly higher frequency, the quanta will have a very slightly higher amount of energy, and so the amount of energy is continuous and not discreet. Only if frequency were discreet would energy be discreet. We do not think that frequency itself is intrinsically quantitised for to do so would be to think that space and time are intrinsically quantitised – look at the mass-dimensions in the aside above. In short, you can have any amount of electromagnetic energy you want, including an irrational number's worth, but it will then come in quanta of a definite (possibly irrational) amount.

Because electromagnetic energy comes in quanta at a particular frequency, we cannot integrate over all frequencies but must do a discreet summation. This means we are rid of the infinities of the Rayleigh-Jeans formula. In fact, based upon the assumption of electromagnetic energy quantitisation at a particular frequency, Planck produced a formula for the energy density of emission from a black body that fits the experimental data. That formula is:

$$u = \frac{8\pi\lambda^2}{c^3}\frac{h\lambda}{e^{\frac{h\lambda}{KT}} - 1} \qquad (15.10)$$

In which λ is the wave-length. The graph above is a plot of this formula. It was by fitting this equation to the available experimental data that Planck calculated the value of Planck's constant, $h = 6.626 \times 10^{-34} JS$.

By assuming that electromagnetic energy at a specific frequency comes in quanta of magnitude determined by $E = vh$, where h is Planck's constant, we can match the theory to the experimentally observed black body spectrum. The cost of doing this is the overthrow of the simplicity of the classical physics view that waves and particles are distinct entities. Planck's equation is:

$$E = vh \tag{15.11}$$

This equation has energy, a particle property, equal to frequency (multiplied by h), a wave property. If an object has energy, it must have frequency. If an object has frequency, it must have energy. Planck's equation is a statement of wave-particle duality.

15.6 THE WAVE-PARTICLE DUALITY OF NATURE

The typical wave is expressed as:

$$\exp\left(-i\left(\omega t - \vec{k}\cdot\vec{x}\right)\right) \tag{15.12}$$

Aside: In matrix notation, this is:

$$\exp\left(\begin{bmatrix} 0 & -\left(\omega t - \vec{k}\cdot\vec{x}\right) \\ \left(\omega t - \vec{k}\cdot\vec{x}\right) & 0 \end{bmatrix}\right)$$
$$= \begin{bmatrix} \cos\left(\omega t - \vec{k}\cdot\vec{x}\right) & -\sin\left(\omega t - \vec{k}\cdot\vec{x}\right) \\ \sin\left(\omega t - \vec{k}\cdot\vec{x}\right) & \cos\left(\omega t - \vec{k}\cdot\vec{x}\right) \end{bmatrix} \tag{15.13}$$

In which, for waves with velocity $c = $ *speed of light*, we have:

$$\omega = |\vec{k}|c \tag{15.14}$$

Where ω is the angular frequency and $\vec{k}$ is the propagation vector. Thus ω and $\vec{k}$ are wave properties. Energy and momentum are particle properties. From special relativity, we have:

$$\frac{E^2}{c^2} = p^2 + m^2c^2 \tag{15.15}$$

Where p is the momentum. If the mass is zero, which it is for electromagnetic waves, we get:

$$p = \frac{E}{c} \qquad (15.16)$$

From (15.3), (15.14), and (15.16), we get:

$$E = \hbar\omega = h\upsilon$$
$$\vec{p} = \hbar\vec{k} = \frac{h\upsilon}{c} = \frac{h}{\lambda} \qquad (15.17)$$

These formulae have particle properties, momentum and energy, on the left-hand side and wave properties, angular frequency and propagation vector, on the right-hand side.

Aside: In the 4-vector notation of special relativity., this would be written as:

$$\begin{bmatrix} E \\ p_x \\ p_y \\ p_z \end{bmatrix} = \hbar \begin{bmatrix} \omega \\ k_x \\ k_y \\ k_z \end{bmatrix} \qquad (15.18)$$

And the exponent of (15.12) is merely $-i$ multiplied by the 4-vector dot product of the displacement 4-vector and the wave 4-vector:

$$\begin{bmatrix} \omega \\ k_x \\ k_y \\ k_z \end{bmatrix} \bullet \begin{bmatrix} t \\ x \\ y \\ z \end{bmatrix} = \omega t - x k_x - y k_y - z k_z \qquad (15.19)$$

Planck's formula combines waves and particles. The formula makes no sense unless the objects it describes have both particle and wave properties. Since the formula resolves the blackbody problem and leads to predictions that match the experimental observations, physicists of the early 20th century, reluctantly, began to take wave-particle duality seriously. They were pushed to accept this by other types of evidence.

15.7 THE PHOTO-ELECTRIC EFFECT

If a beam of high frequency light is shone on to a metal, electrons are emitted from the metal. This is called the photo-electric effect. If the light is of a low frequency that is less than a particular threshold frequency, which varies with the type of metal, then no electrons are emitted from the metal. If we increase the intensity of the low frequency light, there is no change and there are still no electrons emitted from the metal. If we increase the intensity of the high frequency light, there is an increase in the number of electrons emitted from the metal. As we increase the frequency of the light above the threshold frequency, electrons are emitted from the metal with increasing kinetic energies. The energy of the emitted electrons is of the form:

$$E_{emitted} = W + T \qquad (15.20)$$

Where W is the energy of the emitted electrons at the threshold frequency and T is the kinetic energy of the emitted electrons at the high frequency. None of the above can be explained by the classical physics view that light is a wave.

In 1905, Albert Einstein (1879–1955) assumed that light can carry energy and that the energy of the light is in discreet amounts (quanta) whose magnitude depends upon the frequency such that[8]:

$$E = \hbar\omega \qquad (15.21)$$

With this assumption, the photo-electric effect is easily explained. When-ever the frequency of the light is sufficient high, it will have sufficient energy to overcome the attractive potential of the metal for the electron and the electron will absorb that energy and be freed from the metal. At a lesser frequency, the light will have less energy, and, if the frequency is less than the threshold frequency, this will be insufficient energy to dislodge the electron.

Aside: Albert Einstein won a Nobel prize for his explanation of the photoelectric effect. He was not awarded a Nobel prize for either his development of the theory of special relativity or for his monumental development of the theory of general relativity.

[8.] Einstein, A (1905) Annalen der Physik 17 (6) 132–148.

Einstein's photoelectric hypothesis was verified experimentally by R. A. Millikan in 1916[9]. The quantum nature of the photoelectric effect was verified in 1915 by Duane and Hunt[10] and in 1928 by Lawrence and Beams[11].

To Newtonians, energy is a continuous concept associated with particles. To explain the photo-electric effect, we are having to associate discreet quanta of energy with electromagnetic radiation. To the classical physicist of the late 19th century, this was shocking.

The photo-electric effect is concerned with nothing more than the exchange of energy between light and matter. Because there is nothing more than only the exchange of energy between light and matter, this effect cannot be explained by any mechanism that avoids the discreetness of the quanta of energy of the light.

Between them, blackbody radiation and the photo-electric effect show only that the exchange of energy between light and matter is quantitised into discreet "lumps" (quanta); they do not show that light has a particle-like nature. However, Compton scattering does show the particle-like nature of light.

15.8 COMPTON SCATTERING

We begin by assuming that electromagnetic radiation (x-rays) comes in photons that have momentum. We then hit electrons with this radiation and calculate the consequences using Newton's laws of motion for particles. There will be a change in the momentum of the photon after it has hit an electron. That change will depend upon the angle at which the photon collided with the electron. The change in momenta of the photons will correspond to a change in the energy, $E^2 = p^2c^2$, and thus in the frequency of the photons, $E = h\upsilon$. We arrange the directions of the sources of the electrons and the electromagnetic radiation to be fixed. The Newtonian mechanics of particles predicts a change of frequency that corresponds to the angle of collision and thus, from pre-set sources, we will see a

[9.] R.A.Millikan Phys. Rev. 7,355 (1916).

[10.] W. Duane and F.L.Hunt. Phys Rev, 6, 166 (1915).

[11.] E.O.Lawrence and J.W.Beams, phys Rev, 32, 478 (1928).

change of frequency corresponding to the angle at which the photon is scattered from the electron. This change will not depend upon the frequency of the electromagnetic radiation. The actual formula is:

$$\lambda_1 - \lambda_2 = 2\lambda_e \sin^2\left(\frac{\theta}{2}\right) \tag{15.22}$$

Where λ_e is the Compton wavelength of the electron. This result is exactly what was observed in experiments by Compton between 1919 and 1923[12,13]. We therefore take it that the assumption of the particle-like nature of the photons of electromagnetic radiation is a correct assumption. These results cannot be explained if we assume that electromagnetic radiation is wave-like.

Aside: The Compton wavelength of the electron is: 4×10^{-13} Meter.

Today, we are not shocked at the idea that electromagnetic waves have quantitised energy and quantitised momentum; as babies, we read it scrawled as graffiti on the cot headboard, but to physicists of a hundred years ago, it was most discombobulating.

Aside: The American physicist Arthur Holly Compton (1892–1962) was awarded the Nobel prize for physics in 1927 for his discovery of the Compton effect. From the start, he was a key figure in the Manhattan project that developed the first nuclear bombs. It was Compton who headed the National Defense Research Committee that, in May 1941, first foresaw the use of nuclear power as an energy source and the possibility of nuclear bombs.

15.9 DE-BROGLIE AND DAVISSON AND GERMER

In 1923[14], 1924[15], and 1925, Louis de Broglie (1892–1987) proposed that the formulae that associate particle-like quantities with

---

[12.] A.H.Compton Phys. Rev. 21, 207 (1923).

[13.] A.H.Compton Phys Rev. 22 409 (1923).

[14.] Louis de Broglie Comptes Rendus 177, 507 548, 630 (1923).

[15.] L de Broglie. Phil Mag 47 446 (1924).

waves ought to associate wave-like quantities with particles[16]. Where there is a wave, there is a wave equation, and de Broglie proposed that an electron should be described by the de Broglie wave equation:

$$\exp\left(-i\frac{\left(Et - \vec{p}\bullet\vec{x}\right)}{\hbar}\right) \tag{15.23}$$

Which the reader might like to compare with (15.12) and (15.18) and (15.19). Basically, de Broglie said that a material particle moving with momentum p, has associated with it a wave whose wavelength is given by:

$$\lambda = \frac{h}{p} = \frac{h}{\sqrt{2mE}} \tag{15.24}$$

Wherein λ is the wavelength, h is Planck's constant, m is the particle's mass and E is the particle's energy.

Aside: The average energy of particles at absolute temperature T is given by:

$$E_{Average} = \frac{3}{2}k_B T \tag{15.25}$$

Wherein $k_B = 1.3807 \times 10^{-23}$ J K^{-1} is Boltzmann's constant. Using:

$$E = \frac{p^2}{2m} \tag{15.26}$$

We get the momentum of a particle with mass m as:

$$p = \sqrt{3mk_B T} \tag{15.27}$$

Using (15.24) gives:

$$\lambda = \frac{h}{\sqrt{3mk_B T}} \tag{15.28}$$

The de Broglie wavelength of an atom of oxygen at room temperature ($\sim300K$) is approximately 4×10^{11} meters. The de Broglie wavelength of a molecule of DNA at room temperature is approximately 10^{-14} meters.

---

16. L.de Broglie Ann Physik 3, 22 (1925).

In 1927, Clinton Davisson (1881–1958) and Lester Germer (1896–1971) showed that a beam of electrons reflected from the surface of a nickel crystal forms a diffraction pattern[17]. This is analogous to the diffraction patterns from light upon a diffraction grating. The diffraction pattern persists even if only one electron is reflected. Thus, a single electron has wave-like properties associated with it – that was a shocker. The de Broglie equation accurately described the energy of electrons and the width of the diffraction bands of those electrons impinging upon the nickel crystal in the Davisson and Germer experiment – the de Broglie wave equation of particles was verified by experiment.

Aside: Louis Victor Pierre Raymond, 7th duc de Broglie was awarded the Noble prize for physics in 1929 for his work on electron waves. Interestingly, his first degree was in history.

15.10 THE THOMPSON ATOM

We understand that matter is comprised of large numbers of discreet entities called atoms. This was postulated, perhaps not for the first time, over two thousand years ago by the ancient Greek Democritus (460–370 BC). It became accepted in the early 19th century when chemists were able to show that the chemical elements could be combined in definite amounts to form common substances.

The original concept of an atom was that it was a tiny hard ball, like a microscopic billiard ball. Today, we have a very different concept of an atom, but we did not get from hard billiard ball to today's concept in only one leap. The transition required many decades and much experimental work. In the early 1830s, Michael Faraday experimented with electrolysis and concluded that atoms could exist as ions[18] thereby bringing into question the "billiard ball" nature of atoms. This was followed in 1897 by J. J. Thompson's discovery of

---

[17.] C.Davisson & L Germer Nature 119, 558 (1927).
[18.] Michael Faraday: Experimental Researches in Electricity. Series VII. January 1834.

the electron[19]. Thompson built upon this discovery by formulating a theory of the atom known as the Thompson atom or the Thompson model of the atom. Thompson envisaged the atom as being a round "plum pudding" like object in which the electrons were embedded as plums are in a pudding and in which the rest of the atomic material is positively charged, thereby neutralizing the negative electric charge of the electrons. Thompson calculated this "plum pudding" atom to have a radius of approximately 10^{-10} M.

15.11 THE RUTHERFORD ATOM

The Thompson model of the atom did not last long. In 1909, R. A. Millikan (1868–1953) measured the charge of the electron using oil droplets[20]. He improved the measurements a few years later. Alpha particle scattering experiments[21] carried out by Geiger and Marsden[22] and supervised by Ernest Rutherford (1871–1937) between 1909 and 1911 led to the atom being viewed as a tiny but heavy, positively charged nucleus with electrons orbiting around that nucleus in a way similar to the way that planets orbit around the sun. Although the experiments lead to this view, such a model of the atom makes no sense in classical physics. An orbiting electron is an accelerating charged particle, and accelerating charged particles emit electromagnetic radiation thereby losing energy. If the electron is losing energy, it will lose speed and will spiral into the nucleus. Calculation shows that it should do this in 10^{-10} seconds. Electrons do not do this. Atoms are stable. In addition, as the electron loses energy, it will emit radiation in a continuous range of differing frequencies. Balmer had already shown that atoms do not do this.

The model of the atom as having orbiting electrons could not fit into classical physics without some modification. That modification was provided by Niels Bohr as an "ad hoc" set of rules appended to classical physics.

[19.] J.J.Thompson Phil. Mag. 44, 293 (1897).
[20.] R.A.Millikan Phys Mag XIX 6 (1910) pg. 209.
[21.] E. Rutherford Phil Mag. 21 669 (1911).
[22.] H Geiger and E. Marsden Proc Roy. Soc. A 82, 495 (1909).

Aside: The New-Zealand born Ernest Rutherford 1st Baron Rutherford of Nelson was known as the "father of nuclear physics" because of his experiments into nuclear decay. He was awarded the 1908 Nobel prize for chemistry for his work on the disintegration of atomic nuclei.

15.12 THE BOHR ATOM

The reader might have been taught that Maxwell's equations describe all of electromagnetic phenomena; they do not. Maxwell's equations describe all of classical electromagnetism; they do not describe the behavior of electrons. In 1913, Niels Bohr (1885–1962) proposed "ad-hoc" rules be appended to Maxwell's equations of electromagnetism to explain the behavior of electrons in atoms[23]. It is not pretty to take a set of self-consistent and complete equations such as the Maxwell equations and add bits to them, but this was the best that could be done at the time. Bohr proposed amending the Maxwell equations by adding the condition that angular momentum comes in lumps whose magnitude is given by:

$$l = n\hbar \qquad (15.29)$$

Discreet amounts of angular momentum means discreet speeds of the electrons orbiting an atomic nucleus. This is the same as discreet energies of these orbiting electrons. It follows simply. If an orbiting electron can go at only particular speeds, it must have only particular angular momenta. This is definitely different from Newtonian mechanics!

What we have here is classically continuous variables taking only discreet values.

Bohr then allowed electrons to "jump" from one orbital speed (one energy or one angular momentum if you prefer) to a different orbital speed. Even today, we do not really understand an "electron jump" – we do not know how much time it takes, for example. As the electrons jump from one energy (orbit) to another, they emit or absorb radiation with energy equal to the energy difference between

---

[23.] N. Bohr Phil Mag 26, 476 (1913).

the two orbits. This is the discreet emission of radiation discovered by Balmer. The difference in energies is proportional to Planck's constant because the allowed angular momenta are proportional to Planck's constant.

When Bohr applied these "ad hoc" rules to the hydrogen atom, he got the following. By classical physics, the equation of motion of the electron is:

$$e^2 = ma^3\omega^2 \tag{15.30}$$

Where e is the charge of the electron. Bohr's rules require:

$$ma^2\omega = n\hbar \tag{15.31}$$

Combining these gives the orbit radius as, $r = a_n$, where:

$$a_n = \left(\frac{\hbar^2}{me^2}\right)n^2 = a_0 n^2$$

$$\omega = \frac{me^4}{\hbar^3}\frac{1}{n^3} \tag{15.32}$$

$a_0 = 5.29 \times 10^{-11}$ *Meter* is a fundamental length connected to the hydrogen atom known as the Bohr radius. It is the classical radius of the electron orbit that is of the lowest energy in the hydrogen atom. You can see it as being a measure of the classical size of the hydrogen atom.

Aside: Of all the elements, caesium is the largest atom measured to date at 2.5 angstroms (an angstrom is 10^{-10} *M*) diameter and oxygen is the smallest at 0.6 angstroms diameter - it might be that some atoms are slightly larger than caesium. There is little difference in the size of atoms throughout the periodic table, but the alkaline metals (leftmost column of the periodic table) are the largest.

Using the classical physics notion of total energy being the sum of potential energy and kinetic energy leads to:

$$E_n = \left(-\frac{e^2}{2a_0}\right)\frac{1}{n^2} \tag{15.33}$$

$$\omega_{mn} = \frac{e^2}{2\hbar a_0}\left(\frac{1}{m^2} - \frac{1}{n^2}\right)$$

We have discreet energy levels and discreet angular frequencies. Choosing $m = 2$ and $n = 3, 4, 5$ will give the spectral lines of the hydrogen atom as found by Balmer. We have agreement with observation. Appropriate substitutions lead to the energy levels that Bohr calculated for the hydrogen atom as:

$$E_n^{\text{Hydrogen atom}} = -\frac{\hbar^2}{2m}\frac{1}{a_0^2}\frac{1}{n^2} \tag{15.34}$$

These discreet energy levels correspond to electron orbits being "circular" standing waves around the atomic nucleus of wavelength given by $n\lambda = 2\pi r$. A "circular" standing wave, very much like a straight standing wave, must match full wavelengths to the length of the "circumference". So it is that electrons having wavelike properties leads to the discreet energy levels of atoms.

Should we emphasize that last sentence? Perhaps we should just repeat it. Because electrons are waves, electrons in atoms have discreet energy levels corresponding to an integral number of wavelengths of the electrons. We would mention the discreet amounts of orbital angular momentum as well, but the reader is clever enough to realize that anyway.

Aside: The Danish physicist Niels Bohr won the Nobel prize for physics in 1922 for his work on quantum theory. Two years previous to this, in 1920, he founded the Institute of Theoretical Physics at Copenhagen, now known as the Niels Bohr Institute. His mother was Jewish, and in 1943, aided by the Danish resistance, he was forced to flee from Denmark to the neutral country of Sweden where-after he played a significant role in organizing the rescue of some 7,000 Danish Jews from the Nazis into Sweden.

The Bohr theory applies to only circular orbits, but we now know that elliptical orbits are allowed. Bohr's theory was generalized to include elliptical orbits by Sommerfield in 1916[24].

[24.] A. Sommerfield Ann. D. Physik 51, 1 (1916)

15.13 THE STERN-GERLACH EXPERIMENT (SEE LATER CHAPTER)

In 1922, Otto Stern (1888–1969) and Walther Gerlach (1889–1979) conducted an experiment that demonstrated the quantitisation of angular momentum[25]. The experiment is universally known as the Stern-Gerlach experiment. The Stern-Gerlach experiment measured the magnetic dipole moments of neutral silver atoms by passing them through a non-uniform magnetic field. The experiment showed that an atomic dipole moment has only discreet values. This is direct experimental evidence of the quantitisation of angular momentum. However, the angular momentum which is discovered by this experiment is the intrinsic spin of the electron and not the orbital angular momentum originally postulated by Bohr. Still, although it be intrinsic spin, this experiment demonstrated that angular momentum is quantitised.

***

Aside: The results of the Stern-Gerlach experiment were reported directly to Niels Bohr by Walther Gerlach on a post card.

Aside: The velocity of an electron in the lowest Bohr orbit is:

$$v = \frac{e^2}{\hbar} \qquad (15.35)$$

If we divide this by the speed of light, c, we get the fine structure constant:

$$\alpha_{fine-structure} = \frac{v}{c} = \frac{e^2}{\hbar c} = \frac{1}{137} \qquad (15.36)$$

In this, we see three basic physical constants combined. The fine structure constant measures the strength of the electromagnetic interaction. The fine structure constant is also a measure of the speed, $\alpha c = \dfrac{c}{137}$, of the electron in the lowest orbit Bohr atom. It was introduced by Arnold Sommerfeld (1868–1951) in 1916.

***

[25] Gerlach,W: Stern,O: (1922) Das magnetische moment des silberatoms: Zeitschrift für Physik 9. 353–355

SUMMARY

Taken together, the black-body spectrum, the photo-electric effect, Compton scattering, and Bohr's successful explanation of the spectra of the hydrogen atom form a large body of evidence supporting the idea that atomic particles and electromagnetic radiation have both wave-like and particle-like properties. More important is the fact that none of these phenomena can be explained by the classical physics view of a universe that is separated into waves and particles. Today, there is much more evidence of the wave-particle nature of both electromagnetic radiation and atomic particles to be piled on top of the four phenomena listed above. The quantum type of mechanics that accompanies the wave-particle view has been, and continues to be, triumphantly successful in predicting and explaining phenomena (think superfluidity and superconductivity and the big bang synthesis of the elements and Bose-Einstein states and, quantum computing, and ...).

So, by the middle of the 19th century, things were simple. We had particles, and we had waves, and they were never mixed together. By the end of the first quarter of the 20th century, we had particles that were also waves and waves that were also particles. Now, in the 21st century, we still refer to atomic particles and electromagnetic waves just as 19th century physicists would have done - presumably we do this to confuse students - we ought really to have changed the vocabulary, but we haven't.

The change from the 19th century view to the present day view came about in three imposed "ad hoc" rules of thumb:

i. The particle aspect of radiation – Planck's photons (Einstein helped).

ii. The wave aspect of particles – de Broglie's wave equation.

iii. Some physical variables have only a discreet set of values – Bohr.

The first two of these mean that we need a formulation of mechanics that combines both waves and particles. This cannot be done with the Newtonian formulation of mechanics. The last of these means that we need a mathematical formulation that gives a particular set of numbers associated with a particular physical variable like energy or angular momentum. Nor can this be done with the Newtonian formulation of mechanics. The wave-particle duality of objects and the discreet nature of energy or angular momentum is the downfall, for tiny objects, of Newtonian dynamics.

Aside: Although, historically, the quantitisation of energy came before the de Broglie wave equation of particles, it might have happened differently. If de Broglie had proposed that electrons were waves first, then it would have followed the electron orbits are standing waves of a particular whole number of wavelengths and hence angular momentum is quantitised. Hence energy is quantitised, and hence when electromagnetic radiation interacts with atoms it will do so in discreet quanta of energy. These quanta of energy are called photons. As well as delivering energy to orbiting electrons, photons must deliver angular momentum to accommodate the change in electron orbit. It is not a big step to the idea that photons have particle-like momentum.

EXERCISE

1. What is the de Broglie wavelength of a neutron (mass 1.67×10^{-27} Kg) at room temperature (~300 K)?

16

QUANTUM MECHANICS – AN OVERVIEW

In this chapter, we skim through quantum mechanics to give a general overview without too many details. We repeat much of the previously covered material to "ram it home." Some of the previously covered material is presented in a different way to broaden the reader's understanding. Some of the material is yet to be covered, and so this chapter is a preview of that material.

16.1   THE BONES OF THE THEORY OF QUANTUM MECHANICS

1.   a. *Quantum mechanics is based on linear operators:* At the simplest level, a linear operator is no more than a $n \times n$ square matrix that acts upon a n-component vector by matrix multiplication to change it into another vector. In quantum mechanics, we are interested in only the special vectors that are unchanged by a linear operator other than to be multiplied by a real number; we call these eigenvectors.

The number by which the operator multiplies the eigen-vector is called an eigenvalue.

$$
\begin{bmatrix} a & 0 & 1 & -1 \\ 0 & a & 1 & 1 \\ 1 & 1 & a & 0 \\ -1 & 1 & 0 & a \end{bmatrix}
\begin{bmatrix} \dfrac{1}{\sqrt{2}} \\ \dfrac{1}{\sqrt{2}} \\ 1 \\ 0 \end{bmatrix}
=
\left(a + \sqrt{2}\right)
\begin{bmatrix} \dfrac{1}{\sqrt{2}} \\ \dfrac{1}{\sqrt{2}} \\ 1 \\ 0 \end{bmatrix}
\qquad (16.1)
$$

$$\quad\text{Operator}\qquad\quad\text{Eigenvector}\qquad\quad\text{Eigenvalue}\qquad\text{Eigenvector}$$
$$\text{(unchanged)}$$

Eigenvector means "special vector." Eigenvalue means "special value." Eigenfunction means "special function."

b. *The sums and products of linear operators are linear operators:* Because the operators are linear, their sums and products are linear operators.

2. a. *In quantum mechanics, there is one, and only one, linear operator corresponding to each Newtonian dynamic variable (except time):* In Newtonian dynamics, we have dynamic variables like energy, x-momentum, y-momentum, z-angular momentum, etc... In quantum mechanics, for every Newtonian dynamic variable, there is a corresponding linear operator. There is the energy operator; there is the x-momentum operator; there is the y-momentum operator; there is the z-angular momentum operator etc... .

b. *Swapping the Newtonian variables of a physical system for linear operators imposes the commutation relations of those operators upon the physical system.* There are no commutation relations within Newtonian mechanics, and so, when we quantitise the physical system by replacing Newtonian variables with operators, we are fundamentally changing the physical system.

c. *In quantum mechanics, there are operators that have no corresponding dynamic variable in Newtonian mechanics:* Although every Newtonian dynamic variable except time has a corresponding operator in quantum mechanics, there are also other operators in quantum mechanics for which

we do not find a corresponding dynamic variable in Newtonian dynamics. An example of this is the intrinsic spin of
a particle.

Aside: Other examples of operators are found in QFT but not in
quantum mechanics; examples are the creation operator that creates
a particle from energy and the annihilation operator that annihilates
a particle into energy.

---

    d. *The energy operator is called the Hamiltonian:* We often
    refer to the energy operator as the energy operator, but
    we also refer to it as the Hamiltonian operator or just the
    Hamiltonian of the physical system.

3. *In quantum mechanics, the eigenvalues associated with an
operator are the possible values of the Newtonian variable
associated with that operator:* The eigenvalues of the energy
operator are the possible energies of a particular physical system. Sometimes, the set of energy eigenvalues is the whole of
the real numbers. Sometimes, the set of energy eigenvalues
is a finite discreet set of real numbers. The energy eigenvalues of a free particle are the whole of the real numbers, and
so a free particle can have any value of energy (just like a
Newtonian particle). The energy eigenvalues of an electron
in orbit around an atom are a discreet set of real numbers
corresponding to a discreet set of standing waves. The energy
eigenvalues of an electron in orbit around an atom are the
discreet possible energies of that electron and the difference
between these discreet energies correspond to the frequencies of the emitted, or absorbed, spectral lines on that atom.

You see, we have managed to capture both the Newtonian
free particle energies and the atomic quantitised energies
within our operator/eigenvector/eigenvalue system.

4. a. *In quantum mechanics, the eigenvectors associated with
an operator correspond to the possible states in which the
system might be with regard to the Newtonian variable associated with that operator:* An electron orbiting an
atom might be in one of, say, ten possible orbits. In which
case, the energy operator of that system will have ten

eigenvectors. Each energy eigenvector will be associated with one of the possible orbits of the electron. Associated with each energy eigenvector will be one energy eigenvalue. Each energy eigenvalue will be the electron's energy in the associated orbit. It is not necessarily the case that the ten eigenvalues will all be different.

Similarly, there will be ten z-angular momentum eigenvectors, one for each orbit, and ten associated z-angular momentum eigenvalues corresponding to the ten values of z-angular momentum, one for each orbit.

Depending upon the system, the set of ten eigenvectors of the different operators might or might not be the same.

   b. *In quantum mechanics, a particular state of a physical system is called an eigenstate of that system:* Corresponding to a particular orbit of an electron is a particular energy (the energy eigenvalue), and a particular energy eigenfunction.

5. a. *Linear operators can be written without matrices:* In quantum mechanics there is a x-momentum operator which is differentiate with respect to x and multiply by $i^3\hbar = -i\hbar = -\sqrt{-1}\hbar$. The x-momentum operator is written as $\widehat{p_x}$. We have the three quantum mechanical operators corresponding to the three Newtonian dynamic variables:

$$\widehat{p_x} = -i\hbar\frac{\partial}{\partial x}$$
$$\widehat{p_y} = -i\hbar\frac{\partial}{\partial y} \tag{16.2}$$
$$\widehat{p_z} = -i\hbar\frac{\partial}{\partial z}$$

It is normal to put a carat over an operator to signify that we regard it to be an operator. Thus, we write the energy operator as:

$$\widehat{E} \quad \text{or} \quad \widehat{H} \tag{16.3}$$

The practice of calling the energy operator the Hamiltonian is the origin of this doubling of notation.

b. *The time momentum variable is:*

$$\widehat{p_t} = i\hbar \frac{\partial}{\partial t} \tag{16.4}$$

The theory of special relativity tells us that energy is momentum in the time direction. The minus sign is absent in the time momentum operator because the signature of the distance function of 4-dimensional space-time is (+, −, −, −). So what we really have here is the energy operator.

$$\widehat{E} = \widehat{p_t} \tag{16.5}$$

Well, yes we do have one form of the energy operator; there is another, more commonly used, form of the energy operator which we will meet shortly.

Aside: If we write special relativity using the hyperbolic complex numbers, we have the relation:

$$\begin{bmatrix} E & p \\ p & E \end{bmatrix} \equiv \begin{bmatrix} h & 0 \\ 0 & h \end{bmatrix} \begin{bmatrix} \cosh \chi & \sinh \chi \\ \sinh \chi & \cosh \chi \end{bmatrix}$$

$$\begin{bmatrix} 0 & p \\ p & 0 \end{bmatrix}^2 = \begin{bmatrix} p^2 & 0 \\ 0 & p^2 \end{bmatrix} \equiv \begin{bmatrix} E & 0 \\ 0 & E \end{bmatrix} \tag{16.6}$$

In this, we see the relation between energy and momentum $E \sim p^2$.

---

6. *The relations between Newtonian dynamic variables are duplicated between the quantum mechanical operators that correspond to those Newtonian dynamic variables:* Within Newtonian mechanics, we have the zero potential (a uniform potential is a zero potential) energy/momentum relation:

$$E = \frac{1}{2}mv^2 = \frac{1}{2m}p_x{}^2 \tag{16.7}$$

Which we have shown in only one dimension. It is often normal to omit the subscript $\{x, y, z\}$ from the momentum operator when we write 1-dimensional forms of equations. Within quantum mechanics, we duplicate this relationship and we have the relation between the energy operator (the Hamiltonian) and the momentum operator:

$$\widehat{E} = \frac{1}{2m}\widehat{p}^2$$

$$= \frac{\hbar^2}{2m}\left(-i\frac{\partial}{\partial x}\right)\left(-i\frac{\partial}{\partial x}\right) \qquad (16.8)$$

$$= -\frac{\hbar^2}{2m}\frac{\partial^2}{\partial x^2}$$

Aside: The reader might think we derived the energy operator wrongly and that we ought to have done the calculation as:

$$\widehat{E} = \frac{\widehat{1}}{2m}\widehat{p}_x^2$$

$$= -\frac{\hbar^2}{2m}\left(\frac{\partial}{\partial x}\right)^2 \qquad (16.9)$$

We will eventually put the energy operator into a wave equation (the Schrödinger equation). Any respectable wave equation should be linear (the Schrödinger equation is linear) because it is an observed fact that waves can be superimposed upon each other to form another wave. Since the solutions of a wave equation are waves, we want to be able to superimpose the solutions upon each other to form another solution just as waves can be imposed upon each other to form another wave. We can do this with only linear equations. If we had an energy operator with a $\left(\dfrac{\partial}{\partial x}\right)^2$ term, we could not form a linear equation that included it; this would mean that our "wave equation" was not linear and therefore not a respectable wave equation. (It would also mean that the solutions of the equation were not a complete vector space with orthogonal basis solutions.)

To some extent, the choice is arbitrary. The choice is justified, not by any profound understanding of the universe, but by the fact that it produces an equation, the Schrödinger equation, that correctly describes reality. The reader may take the view that the choice is a postulate of quantum mechanics.

We now have another form of the energy operator.

Correspondence Principle

Every Newtonian dynamic variable has a corresponding quantum mechanical operator.

The relations between Newtonian dynamic variables are duplicated as relations between the corresponding quantum mechanical operators.

The quantum mechanical operators bring with them the commutation relations between them and thereby are commutation relations imposed upon the Newtonian system.

Another example of relations between operators being the mirror of the Newtonian relations are the angular momentum operators:

$$\widehat{l}_x = \widehat{y}\widehat{p_z} - \widehat{z}\widehat{p_y}$$
$$\widehat{l}_y = \widehat{z}\widehat{p_x} - \widehat{x}\widehat{p_z} \qquad (16.10)$$
$$\widehat{l}_z = \widehat{x}\widehat{p_y} - \widehat{y}\widehat{p_x}$$

To reiterate, the reader should note that replacing the Newtonian variables by operators has introduced commutation relations to the physical system.

7. *Putting the two forms of the energy operator together forms the time dependent Schrödinger equation (TDSE):* From above, we have two forms of the energy operator. We will give them something to operate upon, Y. Clearly, they are equal to each other because there is only one energy operator, and we have:

$$\widehat{p_t}(\Psi) = \widehat{E}(\Psi)$$
$$i\hbar \frac{\partial \Psi}{\partial t} = -\frac{\hbar^2}{2m}\frac{\partial^2 \Psi}{\partial x^2} \qquad (16.11)$$

This is known as the Time Dependent Schrödinger Equation, abbreviated to TDSE, for a uniform (zero) potential. For a non-uniform potential, we add the potential, $V(x, t)$, as:

$$i\hbar \frac{\partial \Psi}{\partial t} = -\frac{\hbar^2}{2m}\frac{\partial^2 \Psi}{\partial x^2} + V\Psi \qquad (16.12)$$

This is the Time Dependent Schrödinger Equation for a non-uniform potential. It is the backbone of the wave mechanics part of quantum mechanics. We will meet it again later.

Although the Schrödinger Equation can be derived in various ways, one of which we have shown above, in each of these derivations, there is an assumption, or several assumptions. Above, we have assumed the Newtonian energy relation. This means that the Schrödinger Equation is a postulate of quantum mechanics. The "derivations" of it are no more than "eye-opening" attempts to justify it.

Aside: The complex conjugates of the two energy operators are also equal.

$$i\hbar \frac{\partial \psi^*}{\partial t} = -\frac{\hbar^2}{2m} \frac{\partial^2 \psi^*}{\partial x^2} + V(t,x)\psi^* \tag{16.13}$$

Only the $\widehat{E} = -\frac{\hbar^2}{2m} \frac{\partial^2}{\partial x^2} + V$ operator is known as the Hamiltonian, $\widehat{H}$.

8. *The quantum mechanical energy operator is non-relativistic:* Quantum mechanics is a non-relativistic theory because the energy operator is calculated using the non-relativistic energy momentum relation:

$$E = \frac{1}{2m}p^2 \tag{16.14}$$

instead of the relativistic energy/momentum relation:

$$E^2 = p^2c^2 + m^2c^4 \tag{16.15}$$

This does not mean that special relativity is ignored completely in quantum mechanics[1]. We have above derived the TDSE from the relativistic fact that energy is momentum in the time direction.

Aside: The relativistic expression for energy is:

$$E^2 = c^2p^2 + m_0^2c^4 \tag{16.16}$$

---

[1] General relativity is ignored completely in quantum mechanics.

The reader might think it better to use this relation between energy and momentum rather than the Newtonian relation we have used above. The reader would be correct. Relativity theory tells us that energy is momentum in the time direction, and so the energy operator is thus:

$$\widehat{E} = \widehat{p_t} = i\hbar \frac{\widehat{\partial}}{\partial t}$$

(16.17)

Putting this energy operator and the spatial momentum operator into the relativistic energy/momentum relation gives:

$$E^2 = c^2 p^2 + m_0^{\,2} c^4$$

$$i\hbar \overbrace{\frac{\partial}{\partial t}}^{2} (\psi) = -i\hbar \overbrace{\frac{\partial}{\partial x}}^{2} (\psi)(c^2) + m_0^{\,2} c^4 (\psi)$$

(16.18)

Setting $c = 1$ and re-arranging gives:

$$\frac{\partial^2 \psi}{\partial t^2} - \frac{\partial^2 \psi}{\partial x^2} + m_0^{\,2}\psi = 0$$

(16.19)

A change of notation:

$$(\partial_\mu \partial^\mu + m^2)\psi = 0$$

(16.20)

This is the Klein-Gorden equation of quantum field theory, QFT. This is exactly where Schrödinger was initially led. The Klein-Gorden equation describes particles of integral intrinsic spin (scalar fields).

It is because we use the Newtonian energy/momentum relation to build quantum mechanics rather than the relativistic energy/ momentum relation that quantum mechanics is a non-relativistic theory. Quantum mechanics deals with only slow moving physical situations, and so we can "get away" with a non-relativistic theory.

---

9. *The eigenvectors (eigenfunctions) of a quantum mechanical operator form a complete orthogonal set:* The linear sums of eigenvectors of a quantum mechanical operator form a linear space (also called a vector space).

10. *The solutions of a linear differential equation form a vector space:* Of all the solutions of a (any) linear differential equation, some are basis solutions that are mutually independent

of each other (as defined by the overlap integral) and form a complete set. The other solutions of the linear differential equation are formed as a linear sum of the basis solutions. The solutions of non-linear differential equations do not have this orthogonality and completeness. The time dependent Schrödinger equation, TDSE, is a linear differential equation. The basis solutions of the TDSE form a complete orthogonal set. There is nothing special about the TDSE; the solutions of all linear differential equations have this orthogonality and completeness property.

11. *The eigenfunctions (eigenvectors) of an operator vary from one physical situation to another:* The set of eigenfunctions of the energy operator within a non-uniform potential are different from the set of eigenfunctions of the energy operator of a free particle (a particle in a uniform potential). There is only one energy operator but its exact form and its eigenfunctions vary from one physical system to another.

12. *In quantum mechanics, different operators can have the same eigenvectors (eigenfunctions):* It is the case that the zero potential Hamiltonian (the energy operator in a uniform potential) and the momentum operator have the same set of eigenfunctions. These two operators might associate different eigenvalues with the eigenfunctions, but the set of eigenfunctions are the same.

13. *Operators with the same set of eigenfunctions are commutative:* We take the example of the zero potential energy operator and the x-momentum operator. By multiplicatively commutative, we mean that the order in which the operators are applied does not matter. If this is the case, then the difference between the two possible orders of application will be zero. We call this difference the commutator of the operators and, for operators $\{\widehat{A}, \widehat{B}\}$ acting on vector (function) Ψ, we write it as:

$$\left[\widehat{A}, \widehat{B}\right]\Psi = \widehat{A}\left(\widehat{B}(\Psi)\right) - \widehat{B}\left(\widehat{A}(\Psi)\right) \qquad (16.21)$$

Note that the commutator is an operator in its own right. We have:

$$\widehat{E}_{V=0} = -\frac{\hbar^2}{2m}\frac{\partial^2}{\partial x^2}$$

$$\widehat{p}_x = -i\hbar\frac{\partial}{\partial x}$$

$$\left[\widehat{E},\widehat{p}\right]\Psi = -\frac{\hbar^2}{2m}\frac{\partial^2}{\partial x^2}\left(-i\hbar\frac{\partial}{\partial x}(\Psi)\right) - -i\hbar\frac{\partial}{\partial x}\left(-\frac{\hbar^2}{2m}\frac{\partial^2}{\partial x^2}(\Psi)\right)$$

$$= i\frac{\hbar^3}{2m}\frac{\partial^3\Psi}{\partial x^3} - - - i\frac{\hbar^3}{2m}\frac{\partial^3\Psi}{\partial x^3} = 0$$

(16.22)

We see that the zero potential Hamiltonian operator and the momentum operator commute. Not all operators commute, and, indeed, the non-zero potential Hamiltonian operator does not commute with the momentum operator.

14. *If operators commute with each other, then the Newtonian dynamic variables which they each represent can be simultaneously known:* Because the zero potential Hamiltonian operator and the momentum operator have the same set of eigenfunctions (eigenvectors), any eigenfunction will have associated with it two eigenvalues, one for the Hamiltonian operator and one for the momentum operator. We often write these two numbers as $|a,b\rangle$. We can observe both these numbers simultaneously with unlimited accuracy. $\{a, b\}$ are, of course, quantum numbers of the system.

15. a. *If operators do not commute with each other, then the Newtonian dynamic variables which they each represent cannot be simultaneously known:* The state of a system corresponds to a particular eigenfunction; the system is in a state described by only one eigenfunction. The x-position operator and the x-momentum operator have different eigenfunctions. Thus, the system cannot simultaneously be in a state described by a x-position eigenfunction and in a state described by a x-momentum eigenfunction. So, because the system can be in only one state at a time, we cannot simultaneously know an eigenvalue of x-position and an eigenvalue of x-momentum. To do this, we would

need the system to be in two different states described by two different eigenfunctions at the same time.

16. a. There is an uncertainty relation between the position and the momentum of a physical system (particle) which is expressed as:

$$\Delta x \Delta p \geq \frac{1}{2}\hbar \qquad (16.23)$$

We also have the uncertainty relation:

$$\Delta t \Delta E \geq \frac{1}{2}\hbar \qquad (16.24)$$

b. In general, for two operators, $\{\hat{A}, \hat{B}\}$, we have the uncertainty relation:

$$\Delta A \Delta B \geq \frac{\left\langle \left[\hat{A}, \hat{B}\right] \right\rangle}{2i} \qquad (16.25)$$

Where the $\left\langle \left[\hat{A}, \hat{B}\right] \right\rangle$ is the expectation value of the commutator.

17. *There is a formula for converting non-matrix operators into matrix operators:* If an operator has n basis eigenfunctions (basis eigenvectors), then those eigenfunctions form a n-dimensional linear space. Such a n-dimensional linear space corresponds to a $n \times n$ matrix. The elements of that $n \times n$ matrix are each given by an overlap integral of the eigenfunctions. If the n eigenfunctions are $\{\phi_i\} : 1 \leq i \leq n$, then the element of the matrix on the Rth row and in the Cth column is given by the overlap integral:

$$A_{RC} = \int_{-\infty}^{\infty} dx\ \phi_R^{\circ}\phi_c \qquad (16.26)$$

So, an operator with three basis eigenfunctions corresponds to a 3×3 matrix and an operator with an infinite number of basis eigenfunctions corresponds to an $\infty \times \infty$ matrix. It is the size of this $\infty \times \infty$ matrix that causes the notational problem of not being able to explicitly write all operators as matrices.

18. *Much of quantum mechanics is about solving eigenvalue equations:* An eigenvalue equation is simply:

$$\widehat{A}(\psi) = a_n\psi \tag{16.27}$$

$$\overline{\text{Operator (eigenfunction)}} = \text{eigenvalue} \times \text{eigenfunction}$$

We solve eigenvalue equations by finding the set of eigenvalues and eigenfunctions which satisfy the equation.

19. *The time independent Schrödinger equation, TISE, is an eigenvalue equation:* We have the time dependent Schrödinger equation, TDSE is:

$$i\hbar\frac{\partial\psi}{\partial t} = -\frac{\hbar^2}{2m}\frac{\partial^2\psi}{\partial x^2} + V\psi \tag{16.28}$$

If the physical system is independent of time such as an electron in a stable orbit around an atomic nucleus, then the eigenfunctions, y, which correspond to the possible states of that physical system will be independent of time and $\frac{\partial\psi}{\partial t} = 0$.

In this case, we are left with only one form of the energy operator, the Hamiltonian:

$$\widehat{E}(\psi) = -\frac{\hbar^2}{2m}\frac{\partial^2\psi}{\partial x^2} + V\psi \tag{16.29}$$

This is the energy operator acting upon an eigenfunction and it must produce a real multiple of that eigenfunction. Therefore, we have, for physical systems that are independent of time:

$$-\frac{\hbar^2}{2m}\frac{\partial^2\psi(x)}{\partial x^2} + V\psi(x) = E_n\psi(x) \tag{16.30}$$

Where E_n is the energy eigenvalue. This is called the time independent Schrödinger equation, abbreviated to TISE. It is a linear differential equation, and so it has solutions that form a vector space spanned by basis functions. The TISE is an eigenvalue equation wherein the only admissible solutions are eigenfunctions of the energy operator (Hamiltonian). The vector space of solutions is therefore spanned by a set of eigenfunctions. Other solutions of the TISE can be formed as linear sums of these basis eigenfunctions just as other vectors can be formed as linear sums of basis vectors. A particular such sum might be:

$$\Psi = c_1\psi_1 + c_2\psi_2 + c_3\psi_3 + ... \qquad (16.31)$$

Wherein the $c_i \in \mathbb{C}$ are constants which are complex. In general, the eigenfunctions, ψ_i, are complex functions.

We use the TISE to deal with time independent physical systems. We simply[2] insert the appropriate potential, $V(x)$ into the TISE and find the set of eigenfunctions and the set of eigenvalues. The hydrogen atom is a physical system that does not change over time. The hydrogen atom is described by the TISE. The eigenvalues of the solutions of the hydrogen atom TISE are the electron energies of the hydrogen atom.

20. *In quantum mechanics, the universe exists in a superposition of eigenstates:* A physical system is described by the time dependent Schrödinger equation, TDSE. This equation has several basis solutions (basis eigenfunctions) each of which corresponds to a particular eigenvalue and a particular basis state of the system. We do not say that the physical system is in one of the basis states corresponding to one of the basis solutions (basis eigenfunctions); instead, we say that the physical system is in all the states at the same time. We say that the physical system is in a superposition (linear sum) of these states. However, when we measure the system, or the system interacts (think electron in orbit being hit by a photon of light), then the system must respond as if it is in one particular state with only one value of energy and only one value of angular momentum and only one value of momentum... etc. We say that the system has collapsed from its superposition of states into a single state. We call this single state an eigenstate.

Aside: The superposition idea is very like being outside of time. A physical system, say an atom, might be in one physical state on Tuesday and in a different physical state on Wednesday and in yet another physical state on Thursday. Now take away time; in which physical state is the atom? Now imagine that a physics student who

---

[2.] Although inserting the potential might be simple, solving the equation is usually not that simple.

lives in space-time comes along to observe the atom. His observation introduces time to the atomic system.

21. *Chance determines into which basis eigenfunction of which operator a system will collapse:* It is perhaps the most mind blowing aspect of quantum mechanics that probability plays a role in the unfolding of events. When a physical system collapses into a particular basis eigenfunction (particular basis solution of the eigenvalue equation), how does it choose the "chosen eigenfunction"? Remarkably, it doesn't; there is no "hidden variable" that determines which eigenfunction will be the "chosen one." It is all down to chance.

22. *The probability of an eigenfunction being the "chosen one" when a physical system collapses into an eigenstate is determined by the complex coefficient of that eigenfunction within the linear sum of eigenfunctions that is the superposition of eigenfunctions that describes the physical system:* The superposition of eigenfunctions is:

$$\Psi = c_1\psi_1 + c_2\psi_2 + c_3\psi_3 + \dots \qquad (16.32)$$

The size (modulus) of the complex coefficient, c_i, of the eigenfunction, y_i, determines how similar this eigenfunction is to the physical system – remember, all eigenfunctions are dissimilar (orthogonal) from each other. The normalized size of the complex coefficient is the modulus of that coefficient. This is proportional to the probability of the particular eigenfunction being the "chosen one".

$$P \propto |c_i|^2 = c_i^* c_i \qquad (16.33)$$

$c_i^* c_i$ is just the "length" of the complex number – its distance from the origin of the complex plane.

23. *Any operator that commutes with the Hamiltonian is associated with a Newtonian variable that is conserved:* The conservation laws of quantum mechanics, like conservation of angular momentum, are based upon whether or not the particular operator commutes with the energy operator – see Ehrenfest's theorem.

24. *Angular momentum comes in two types known respectively as orbital angular momentum and intrinsic spin:* It is one of the great surprises of quantum mechanics that, in addition to the Newtonian orbital angular momentum, we have another type of angular momentum (called intrinsic spin) which has no classical counterpart. The orbital angular momentum operators are odd sized square matrices, and the spin operators are even sized matrices.

25. *There are two types of particles in the universe, fermions and bosons:* Some particles have half integral amounts of intrinsic spin; these are called fermions. Electrons and quarks are fermions. Other particles have integral amounts of intrinsic spin; these are called bosons. Photons are bosons. Fermions obey the Pauli exclusion principle. Bosons do not obey the Pauli exclusion principle.

CHAPTER

17

BASIC FORMULAE

17.1 SOME FORMULAE

1. The energy, E, of a photon of electromagnetic radiation (light, radio waves) is the frequency, v, of that electromagnetic radiation multiplied by Planck's constant, h:

$$E = vh \qquad (17.1)$$

Or

$$E = \omega\hbar \qquad (17.2)$$

Where $\hbar = \dfrac{h}{2\pi}$, $\omega = 2\pi v$

2. The relativistic energy relation between energy, E, and momentum, p, is:

$$E^2 = p^2c^2 + m^2c^4 \qquad (17.3)$$

Wherein c is the velocity of light. For a massless particle, this is:

$$E = pc \qquad (17.4)$$

Using (17.1), this leads to:

$$p = \frac{h}{\lambda} = \hbar k \qquad (17.5)$$

Wherein λ is the wavelength and k is the wave number.

3. The time dependent Schrödinger equation, TDSE, in three spatial dimensions is:

$$i\hbar \frac{\partial \psi(t,x,y,z)}{\partial t} = -\frac{\hbar^2}{2m} \nabla^2 \psi(t,x,y,z) + V(t,x,y,z)\psi(t,x,y,z) \quad (17.6)$$

In one spatial dimension, this is:

$$i\hbar \frac{\partial \psi(t,x)}{\partial t} = -\frac{\hbar^2}{2m} \frac{\partial^2 \psi(t,x)}{\partial x^2} + V(t,x)\psi(t,x) \qquad (17.7)$$

Wherein $\psi(t, x)$ is a time dependent energy eigenfunction and $V(t, x)$ is a potential that varies with time and spatial position. The solutions of this equation, $\psi(t, x)$, are most often called basis wavefunctions or just wavefunctions.

4. The time independent Schrödinger equation, TISE, in three spatial dimensions is the energy eigenvalue equation:

$$-\frac{\hbar^2}{2m} \nabla^2 \psi(x,y,z) + V(x,y,z)\psi(x,y,z) = E_n \psi(x,y,z) \qquad (17.8)$$

In one spatial dimension, this is:

$$-\frac{\hbar^2}{2m} \frac{\partial^2 \psi(x)}{\partial x^2} + V(x)\psi(x) = E_n \psi(x) \qquad (17.9)$$

Wherein E_n is an energy eigenvalue and $\psi(x)$ is an energy eigenfunction. The solutions of this equation, $\psi(x)$, are most often called basis wavefunctions or just wavefunctions. The solutions vary with the nature of the time independent potential, $V(x)$. We use the TISE to deal with potentials that are independent of time.

5. The time independent potential, $V(x)$ of a simple harmonic oscillator is:

$$V(x) = \frac{1}{2} K x^2 \qquad (17.10)$$

Wherein K is a constant.

6. The time independent potential, $V(x)$ of a hydrogen atom is:

$$V(r) = \frac{e^2}{4\pi\varepsilon_0 r} \qquad (17.11)$$

Wherein e is the charge of the electron.

7. The complete wavefunction (superposition of basis wavefunctions) is given by the linear sum of the basis wavefunctions:

$$\Psi = c_1\psi_1 + c_2\psi_2 + c_3\psi_3 + \ldots$$
$$c_i \in \mathbb{C} \qquad (17.12)$$

8. The inner product (overlap integral) of two wavefunctions (they need not be basis wavefunctions) is given by:

$$\langle \varphi, \psi \rangle = \langle \varphi | \psi \rangle = \int_{-\infty}^{\infty} d\tau \; \varphi^\circ \psi \qquad (17.13)$$

We have introduced two ways of writing the inner product. We have:

$$\langle \varphi | \psi \rangle^* = \int_{-\infty}^{\infty} d\tau \left(\varphi^\circ\right)^* \psi^* = \int_{-\infty}^{\infty} d\tau \; \psi^*\varphi = \langle \psi | \varphi \rangle^* \qquad (17.14)$$

9. The basis wavefunctions are orthogonal as defined by the overlap integral. Assuming the basis wavefunctions are normalized, this is expressed as:

$$\int_{-\infty}^{\infty} d\tau \; \psi_m^*\psi_n = \delta_{mn} \qquad (9.15)$$

Wherein δ_{mn} is the Kronecker delta with $\{m, n\}$ being integers. The Kronecker delta is zero if $m \neq n$ and unity if $m = n$. From this, we see:

$$\langle c_i\psi_i | \Psi \rangle = c_i \qquad (17.16)$$

We see that the inner product of a basis wavefunction with the complete wavefunction is the coefficient of the basis wavefunction.

10. a. The two forms of energy operator are:

$$\widehat{H} = -\frac{\hbar^2}{2m}\frac{\partial^2}{\partial x^2} + V \qquad (17.17)$$

$$\widehat{E} = i\hbar\frac{\partial}{\partial t}$$

b. The zero potential energy eigenfunctions are:

$$Ae^{i\frac{E}{\hbar\sqrt{2m}}x} \tag{17.18}$$

11. a. The momentum operators are:

$$\widehat{p_x} = -i\hbar\frac{\partial}{\partial x}, \qquad \widehat{p_y} = -i\hbar\frac{\partial}{\partial y},$$

$$\widehat{p_z} = -i\hbar\frac{\partial}{\partial z}, \qquad \widehat{p_t} = i\hbar\frac{\partial}{\partial t} \tag{17.19}$$

b. The momentum eigenfunctions are of the form:

$$\psi_{mom} = Ae^{i\frac{p_x}{\hbar}x} \tag{17.20}$$

Within which, the momentum eigenvalues are p_x. We note that these eigenfunctions are the same as the eigenfunctions of the zero potential energy operator (17.18).

12. a. Operators, $\widehat{A}$, in functional form, are Hermitian if:

$$\int_{-\infty}^{\infty} d\tau\, \psi^*\widehat{A}(\phi) = \int_{-\infty}^{\infty} d\tau\, \phi\left(\widehat{A}(\psi)\right)^* \tag{17.21}$$

b. Operators in matrix form are Hermitian if they are equal to their conjugate transpose:

$$\widehat{A} = \widehat{A^{*T}} \tag{17.22}$$

c. There is an equivalence between Hermitian matrix operators and Hermitian functional operators given by:

$$A_{mn} = \int d\tau\, \phi_m^*\widehat{A}(\phi_n) \tag{17.23}$$

Wherein the A_{mn} are the elements of the matrix and the ϕ_i are the basis eigenfunctions. Note that this leads to

$A_{mn} = A_{nm}^*$ which is just another way of writing (17.22).

13. The expectation value, $\langle\widehat{A}\rangle$, of the observable associated with an operator $\widehat{A}$ is given by:

$$\langle\widehat{A}\rangle = \int_{-\infty}^{\infty} d\tau\, \Psi^*\widehat{A}(\Psi) \tag{17.24}$$

This is the inner product of Ψ and $\widehat{A}(\Psi)$. Wherein Ψ is a general wavefunction (not a particular eigenfunction). If the system is in a state described by a particular basis eigenfunction, ψ_n, then the expectation value is the eigenvalue, λ_n, associated with that eigenfunction:

$$\langle\widehat{A}\rangle = \langle\psi_n|\widehat{A}(\psi_n)\rangle = \langle\psi_n|\lambda_n\psi_n\rangle = \lambda_n\langle\psi_n|\psi_n\rangle = \lambda_n \qquad (17.25)$$

If the system is in a state described by a linear sum of basis eigenfunction, Ψ, then the expectation value is the weighted average of the eigenvalues of the basis eigenfunctions.

14. The commutator of two operators is:

$$[\widehat{A},\widehat{B}]\psi = \widehat{A}(\widehat{B}(\psi)) - \widehat{B}(\widehat{A}(\psi)) \qquad (17.26)$$

The commutator of two quantum mechanical operators is a multiple of $\hbar$. For example, we have the commutator of the x-position operator, $\widehat{x}$, , and the x-momentum operator, $\widehat{p_x}$:

$$[\widehat{x},\widehat{p_x}] = i\hbar \qquad (17.27)$$

And we have the commutator of the y-position operator, $\widehat{y}$, and the x-momentum operator, $\widehat{p_x}$:

$$[\widehat{y},\widehat{p_x}] = 0 \qquad (17.28)$$

15. The angular momentum operators are mirrors of the Newtonian $L = r \times p$:

$$\widehat{L_x} = \widehat{y}\widehat{p_z} - \widehat{z}\widehat{p_y} = -i\hbar\left(y\frac{\partial}{\partial z} - z\frac{\partial}{\partial y}\right)$$

$$\widehat{L_y} = \widehat{z}\widehat{p_x} - \widehat{x}\widehat{p_z} = -i\hbar\left(z\frac{\partial}{\partial x} - x\frac{\partial}{\partial z}\right) \qquad (17.29)$$

$$\widehat{L_z} = \widehat{x}\widehat{p_y} - \widehat{y}\widehat{p_x} = -i\hbar\left(x\frac{\partial}{\partial y} - y\frac{\partial}{\partial x}\right)$$

16. The probability that an observation of the dynamic variable associated with the operator $\widehat{A}$ will yield the eigenvalue a_n as being the value of that dynamic variable is proportional to

the similarity of the basis eigenfunction, ψ_n, associated with the eigenvalue a_n and the complete wavefunction, Ψ. This is measured by the size (modulus) of the coefficient, c_n of the basis eigenfunction ψ_n within the linear sum of basis eigenfunctions that is the complete wavefunction. We have:

$$\text{Prob} \propto \left| \int_{-\infty}^{\infty} d\tau \; \psi_n^* \Psi \right|^2 \qquad (17.30)$$

We need to normalize the wavefunctions to ensure that total probability is equal to unity. We can see probability as being the same thing as intensity.

17.2   A LIST OF QUANTUM MECHANICAL OPERATORS

The position and time operators are multiply by the variable operators. They have mass-dimension of length[1] $[L]$:

$$\hat{x} = x, \qquad \hat{y} = y, \qquad \hat{z} = z \qquad (17.31)$$
$$\hat{t} = t$$

The momentum operators are differentiate with respect to the imaginary variable; if this is done using the scaled complex numbers, we get the physical constant, $\hbar = \dfrac{1}{\lambda}$. They have mass-dimension of mass $[M]$:

$$\widehat{p_x} = -i\hbar \frac{\partial}{\partial x}, \qquad \widehat{p_y} = -i\hbar \frac{\partial}{\partial y}, \qquad \widehat{p_z} = -i\hbar \frac{\partial}{\partial z} \qquad (17.32)$$
$$\widehat{p_t} = i\hbar \frac{\partial}{\partial t}$$

The momentum operators acting in an electromagnetic field described by the vector potential, $\vec{A}$ are as above but with an added

---

[1] From special relativity, the mass-dimension of time is length.

term to account for the electromagnetic field. They have mass dimension of mass $[M]$. We give only the x-operator:

$$\widehat{p_x^{Emag}} = -i\hbar\frac{\partial}{\partial x} - A_x q \tag{17.33}$$

The kinetic energy operators are based upon $K.E. = \dfrac{p^2}{2m}$. They have mass-dimension $[M]$:

$$\widehat{E_x} = -\frac{\hbar^2}{2m}\frac{\partial^2}{\partial x^2}, \qquad \widehat{E_y} = -\frac{\hbar^2}{2m}\frac{\partial^2}{\partial y^2}, \qquad \widehat{E_z} = -\frac{\hbar^2}{2m}\frac{\partial^2}{\partial z^2} \tag{17.34}$$

The potential energy operator is:

$$\widehat{V} = V \tag{17.35}$$

The energy operators in a non-uniform potential are the kinetic energy operators with an added term. We give only the x-operator:

$$\widehat{H_x} = -\frac{\hbar^2}{2m}\frac{\partial^2}{\partial x^2} + V \tag{17.36}$$

In an electromagnetic field, the energy operators are based upon $K.E. = \dfrac{p^2}{2m}$ with an added term. They have mass-dimension $[M]$. We give only the x-operator:

$$\widehat{E_x^{Emag}} = \frac{1}{2m}\left(-i\hbar\frac{\partial}{\partial x} - A_x q\right)^2 \tag{17.37}$$

The orbital angular momentum operators are based upon $\vec{L} = \vec{r} \times \vec{p}$. They have mass-dimension $[M][L]$. They are:

$$\widehat{L_x} = -i\hbar\left(y\frac{\partial}{\partial z} - z\frac{\partial}{\partial y}\right), \qquad \widehat{L_y} = -i\hbar\left(z\frac{\partial}{\partial x} - x\frac{\partial}{\partial z}\right)$$

$$\widehat{L_z} = -i\hbar\left(x\frac{\partial}{\partial y} - y\frac{\partial}{\partial x}\right) \tag{17.38}$$

The spin angular momentum operators are the Pauli matrices. They have mass-dimension $[M][L]$. They are:

$$\widehat{S_x} = \frac{\hbar}{2}\begin{bmatrix} 0 & 1 \\ 1 & 0 \end{bmatrix}, \qquad \widehat{S_y} = \frac{\hbar}{2}\begin{bmatrix} 0 & -i \\ i & 0 \end{bmatrix}$$

$$\widehat{S_z} = \frac{\hbar}{2}\begin{bmatrix} 1 & 0 \\ 0 & -1 \end{bmatrix} \tag{17.39}$$

The total angular momentum operators are just the sums of the orbital angular momentum operators and the spin angular momentum operators. They have mass-dimension $[M][L]$:

$$\widehat{J}_x = \widehat{L}_x + \widehat{S}_x \tag{17.40}$$

The total orbital angular momentum is:

$$\widehat{L}^2 = \widehat{L}_x \cdot \widehat{L}_x + \widehat{L}_y \cdot \widehat{L}_y + \widehat{L}_z \cdot \widehat{L}_z \tag{17.41}$$

The reader should note that because the operators are linear, they can be defined in any number of dimensions.

CHAPTER

18

DE BROGLIE WAVES LEAD TO THE MOMENTUM OPERATOR

A wave has the mathematical expression:

$$
e^{-i(\omega t - \vec{k}\bullet\vec{x})} = \exp\left[\begin{array}{cc} 0 & -(\omega t - \vec{k}\bullet\vec{x}) \\ \omega t - \vec{k}\bullet\vec{x} & 0 \end{array}\right]
$$
$$
= \left[\begin{array}{cc} \cos(\omega t - \vec{k}\bullet\vec{x}) & -\sin(\omega t - \vec{k}\bullet\vec{x}) \\ \sin(\omega t - \vec{k}\bullet\vec{x}) & \cos(\omega t - \vec{k}\bullet\vec{x}) \end{array}\right] \quad (18.1)
$$

In 1925 de Broglie[1] postulated that particles of matter, like electrons, have a wave-like nature and that this wave-like nature is expressed as:

$$
e^{-i\left(\frac{Et - \vec{p}\bullet\vec{x}}{\hbar}\right)} = \left[\begin{array}{cc} \cos\left(\dfrac{Et - \vec{p}\bullet\vec{x}}{\hbar}\right) & -\sin\left(\dfrac{Et - \vec{p}\bullet\vec{x}}{\hbar}\right) \\ \sin\left(\dfrac{Et - \vec{p}\bullet\vec{x}}{\hbar}\right) & \cos\left(\dfrac{Et - \vec{p}\bullet\vec{x}}{\hbar}\right) \end{array}\right] \quad (18.2)
$$

[1] L. de Broglie Ann Phys. 3, 22 (1925).

What de Broglie did here was to place the "particle" variable energy in the place of the angular frequency and the "particle" variable momentum in place of the propagation vector. This equation accurately described the energy of electrons and the width of the diffraction bands of those electrons impinging upon a nickel crystal in the experiment of Davisson and Germer[2]. Because the de Broglie wave equation for electrons fits with the experimental data, we think it is a correct description of electrons. It is the experimental success of the de Broglie equation that forces us to accept quantum mechanics[3].

Special relativity tells us that momentum is a space thing, (energy is momentum in the time direction) and so, in pursuit of the momentum operator, we have no interest in the part of the de Broglie equation that depends on time. We therefore ignore the time part. We assume that the momentum, $\vec{p}$, is going to be an eigenvalue. This means we need to forget the vector aspect of $\vec{p}$ and think of it as a single real number, p. Technically, we are realigning our axes so that the momentum vector has only one non-zero component. We are left with:

$$e^{i\frac{p}{\hbar}x} = \begin{bmatrix} \cos\left(\frac{p}{\hbar}x\right) & \sin\left(\frac{p}{\hbar}x\right) \\ -\sin\left(\frac{p}{\hbar}x\right) & \cos\left(\frac{p}{\hbar}x\right) \end{bmatrix} \tag{18.3}$$

We now need to search for the operator of which this is an eigenfunction. Comparing this with the eigenfunction of the "differentiate with respect to an imaginary variable" operator of the division algebra $\mathbb{C}$, we have:

$$\begin{bmatrix} \cos\left(\frac{p}{\hbar}x\right) & \sin\left(\frac{p}{\hbar}x\right) \\ -\sin\left(\frac{p}{\hbar}x\right) & \cos\left(\frac{p}{\hbar}x\right) \end{bmatrix} \equiv \begin{bmatrix} \cos\left(\frac{n\theta}{\hbar}\right) & \sin\left(\frac{n\theta}{\hbar}\right) \\ -\sin\left(\frac{n\theta}{\hbar}\right) & \cos\left(\frac{n\theta}{\hbar}\right) \end{bmatrix}$$

$$\theta \equiv x$$

$$\frac{n}{\hbar} \equiv \frac{p}{\hbar} \tag{18.4}$$

---

[2]. C Davisson & L Germer Nature 119, 558 (1927).
[3]. Along with a million other reasons.

This is an eigenfunction of the operator:

$$-i\hbar\frac{\partial}{\partial x} \tag{18.5}$$

Watch:

$$\widetilde{-i\hbar\frac{\partial}{\partial x}}\left(e^{i\frac{p}{\hbar}x}\right) = -ii\hbar\frac{p}{\hbar}e^{i\frac{p}{\hbar}x} = pe^{i\frac{p}{\hbar}x} \tag{18.6}$$

The eigenvalues of these eigenfunctions are $p = n\hbar$; the eigenvalues are the allowed momenta. Since the eigenvalues are the allowed momenta, we call the operator $-i\hbar\frac{\partial}{\partial x}$ the momentum operator, and we call the eigenfunctions of this operator the momentum eigenfunctions. We now have it that the momentum operator is:

$$\hat{p} = -i\hbar\frac{\partial}{\partial x} \tag{18.7}$$

And that the momentum eigenfunctions are of the form:

$$e^{i\frac{p}{\hbar}x} \tag{18.8}$$

and that the allowed values of the momentum are the eigenvalues of this operator.

The space part of the de Broglie waves are eigenfunctions of the momentum operator.

To equate the momentum operator with the "differentiate with respect to an imaginary variable" operator of the complex number division algebra, we have accepted the de Broglie idea that particles are described by a wave equation (18.2) and that the world is described by operators, eigenfunctions, and eigenvalues.

The time independent de Broglie wave equation (with $t = 0$) is an eigenfunction of the momentum operator. Waves that are independent of time are called standing waves.

We see that momentum is proportional to the rate of rotation of the complex number $e^{i\frac{px}{\hbar}}$; it comes in lumps corresponding to complete rotations in the complex plane.

If we had known that momentum was to do with the rate of rotation in the complex plane, we would have been led to the de Broglie equation.

Aside: The above definition of the momentum operator is commonly the only definition considered within an introductory text book to quantum mechanics. However, if the particle to which it relates is electrically charged with charge q and in a 4-vector electromagnetic field, $\left[\varphi, \vec{A}\right]$, the momentum operator is:

$$\hat{p} = -i\hbar\nabla - q\vec{A} \qquad (18.9)$$

19

THE HAMILTONIAN AND SCHRÖDINGER'S EQUATIONS

Since all the quantum mechanical operators must "morph" into Newtonian dynamic variables as we move from microscopic to macroscopic, the relations between the Newtonian variables must be the same as the relations between the quantum mechanical operators. Having got the momentum operator, we can form the energy operator by copying the Newtonian form of the energy.

Within quantum mechanics, the energy operator is known as the Hamiltonian and is denoted by $\widehat{H}$.

$$E = KE + PE = \frac{1}{2m}p^2 + V(x)$$

$$\widehat{H} = \frac{1}{2m}\hat{p}^2 + V(x)$$

$$= \frac{1}{2m}\left(-i\hbar\frac{\partial}{\partial x}\right)^2 + V(x) \tag{19.1}$$

$$= -\frac{1}{2m}\hbar^2\frac{\partial^2}{\partial x^2} + V(x)$$

These operators must act on a function, which we call $\psi(x)$, and so we get:

$$\widehat{H}(\psi) = -\frac{1}{2m}\hbar^2\frac{\partial^2\psi}{\partial x^2} + V(x)\psi \tag{19.2}$$

We seek the eigenfunctions and eigenvalues of the Hamiltonian. The Hamiltonian eigenvalue equation is:

$$\widehat{H}(\psi) = E_n\psi$$

$$-\frac{1}{2m}\hbar^2\frac{\partial^2\psi}{\partial x^2} + V(x)\psi = E_n\psi$$

(19.3)

This last equation is known as time independent Schrödinger's equation, TISE. It was first presented to the world in 1926 by Erwin Schrödinger[1]. The time independent Schrödinger's equation is no more than the eigenvalue equation of the energy operator. The TISE describes the behavior of a particle with energy, E_n, in a potential $V(x)$ that is unvarying over time.

In general, to get the particular quantum mechanical Hamiltonian of a physical system, we write down the classical Hamiltonian expression for the particular physical system and replace the classical variables with the appropriate quantum mechanical operators. If a physical system is governed by the Hamiltonian $H(q_1, q_2, \ldots p_1, p_2 \ldots)$, then the TISE of that physical system is:

$$H(q_1, q_2, \ldots p_1, p_2 \ldots)\psi_n = E_n\psi_n$$

(19.4)

For example, if there are two particles in the physical system, then the TISE is:

$$-\frac{\hbar^2}{2m_1}\frac{\partial^2\psi}{\partial x^2} - \frac{\hbar^2}{2m_2}\frac{\partial^2\psi}{\partial x^2} + V(x_1, x_2)\psi = E_n\psi_n$$

(19.5)

EXERCISE

1. If the classical Hamiltonian of a physical system is:

$$H = \frac{p^2}{2m} + \hbar.t.L_x$$

(19.6)

What is the quantum mechanical Hamiltonian operator, $\widehat{H}$?

---

[1] E. Schrödinger. Ann d Physik 79, 361, 409 (1926).

20

BOUNDARY CONDITIONS ON EIGENFUNCTIONS

It might be the case that the physical system we are considering has boundary conditions associated with it that must be satisfied. This means that the eigenfunctions of the operators in the physical system have those same boundary conditions that must be satisfied. Consider the above momentum operator, and accept that the eigenfunctions of this operator, $u_n(x)$ are restricted by the boundary conditions that $u_n(x)$ must be periodic over the distance, L. Because of the imposition of boundary conditions, these eigenfunctions are different from the same eigenfunctions without the boundary conditions, or with different boundary conditions. The definition of a function includes the boundary conditions imposed on to that function. The basic momentum operator eigenvalue equation is:

$$\hat{p}(u_n) = -i\hbar \frac{\partial}{\partial x}(u_n) = a_n u_n \tag{20.1}$$

And we seek to find the functions, u_n, that satisfy this equation. Consider:

$$u_n = e^{i\frac{a_n}{\hbar}x}$$

$$\frac{\partial}{\partial x}u_n = i\frac{a_n}{\hbar}e^{i\frac{a_n}{\hbar}x} \tag{20.2}$$

$$\hat{p}\left(e^{ia_n x}\right) = -ii\frac{\hbar}{\hbar}a_n e^{ia_n x} = a_n e^{ia_n x}$$

Thus, an eigenfunction of the $\hat{p}$, operator is $u_n = e^{i\frac{a_n}{\hbar}x}$; this is an eigenfunction with no boundary conditions. Without any boundary conditions, the eigenvalues, a_n, of this eigenfunction are the continuous set of all real numbers. However, we specified that the physical system be subject to boundary conditions. The physical system is described by eigenfunctions, and so the eigenfunctions are subject to the boundary conditions. Those boundary conditions are that the eigenfunction be periodic over the distance, L. We therefore require:

$$e^{i\frac{a_n}{\hbar}x} = e^{i\frac{2\pi}{\hbar L}nx}$$

$$a_n = \frac{2\pi}{L}n$$

(20.3)

We see that imposing the boundary condition has reduced the possible eigenvalues from the whole of the real numbers to a discreet set of those real numbers. The particular values of those real numbers is determined by the value of L. As $L \to \infty$, the eigenvalues tend to the whole of the real numbers again.

The above is an exact parallel of the situation of electrons in, or not in, atoms. When an electron is free from any atom, it can have any value of momentum. When the electron is confined within an atom, it is subject to boundary conditions that reflect the fact that an electron's orbit must be an exact number of electron wavelengths in circumference. An electron confined within an atom can have only discreet values of momentum – just like Niels Bohr postulated.

20.1 EIGENFUNCTIONS OF THE MOMENTUM OPERATOR

We might as well repeat the momentum eigenfunctions. The reader will see these so often that she might as well learn them by heart.

The eigenfunctions of the momentum operator are:

$$u_p(x) = Ce^{i\frac{px}{\hbar}}$$

(20.4)

Note that we can include a real constant, C. Such a real constant can be included in every eigenfunction of every operator. Adjusting it allows us to rescale the eigenfunctions to normalize the probabilities that we calculate from the eigenfunctions.

CHAPTER 21

WAVE MECHANICS - THE SCHRÖDINGER EQUATION

We take it that the reader is still sitting comfortably, and so we will proceed. We have now sufficient understanding to tackle the Schrödinger wave equation. The Schrödinger wave equation is the backbone of quantum mechanics. We will do with it what we did with the Newtonian wave equation many chapters ago.

In quantum mechanics, all the information that describes the state of a physical system (a particle perhaps) is encoded in a wavefunction, usually denoted as $\Psi(t, x)$. The equation that determines the evolution in time of this wavefunction is called the time dependent Schrödinger equation, often abbreviated to TDSE.

Aside: Erwin Schrödinger (1887–1961) was born in Vienna and later moved to Germany. He is most famous for the equation named after him that he presented to the world in 1926[1] and for which he won the Nobel prize in 1933, but he is also famous for the "Schrödinger cat" which can be both alive and dead at the same time[2]. He left

---

[1.] E. Schrödinger. Quantitisation as an eigenstate problem. Annalen der Physik (Jan 1929).

[2.] E. Schrödinger. The present situation in quantum mechanics. Trans by: D. Trimmer in Proceedings of the American Philosophical Society (1935).

Germany in 1933 because of his dislike for the Nazi's anti-Semitism, but, in spite of being a Nobel laureate, he was forced to return some few years later because he could not secure an academic position outside of Germany. Both Oxford University in England and Princeton University in the USA turned him down.

Many authors take the wave nature of particles to be a basic postulate of quantum mechanics because we observe that particles have wave-like properties. If a particle has a wave-like nature, there should be a wave equation associated with it, and so we need to find that wave equation.

We begin our search for the particle wave equation by assuming that a particle has associated with it a wavefunction, ψ, which is a mathematical description of a wave. We further assume that the wavefunction determines everything that can be known about the particle. For this to make sense, we need the wavefunction to have specific properties.

We assume that the wavefunction is a single valued function of the space and time co-ordinates, $\psi(t, x, y, z)$. This is assumed with the hindsight that probability is involved in quantum mechanics and the probability, $|\Psi|^2$, can be only one value at any particular point in space-time. Of course, this allows two values of the wavefunction, $\pm\Psi$. We assume that the wavefunction is continuous through space-time. The basis of this is that probability should be defined everywhere. Further, if it is a wavefunction, it should be continuous. We specifically assume, again with hindsight, that the wavefunction, ψ, and the first derivative of the wavefunction, $\dfrac{\partial \psi}{\partial x^i}$, is continuous everywhere. If this were not so, then the second derivative of the wavefunction, $\dfrac{\partial^2 \psi}{\partial x^{i2}}$, would not satisfy the Schrödinger equation for a finite potential.

We assume that the wavefunction is a complex function; there is nothing like this in Newtonian mechanics. We are assuming that a mathematical object that describes a real physical phenomenon is a complex number, $\mathbb{C}$. As we have pointed out earlier, wave equations are all tied to the complex numbers because the "wave" trigonometric functions {cos( ), sin( )} exist in the complex numbers. Since we observe matter to have wavelike properties, these properties are best described using complex numbers.

Within wave mechanics, $\psi^*\psi = |\psi|^2$ is the intensity of the wave. We refer to this as a probability density, but the reader might prefer to think of it as an intensity density. The wave is more intense (there is more of it) in places where $|\psi|^2$ is great and less intense (there is less of it) in places where $|\psi|^2$ is small. We identify intensity with the probability of the wave being found to be at the particular point in space – the more of the wave that there is at a point, the more likely it is to be found there. In order to identify wave intensity with probability density, we have to divide the intensity at every point by an appropriate real number so that the total intensity (= probability) is unity because total probability is unity. This dividing of the intensity at every point by an appropriate real number is called normalization. We have to do it because statisticians normalized probability to unity centuries ago.

We assume that the wavefunction is differentiable. We need this to be the case because we are going to differentiate it to get a wave equation. We assume that the wavefunction is square integrable. The reason for this is that total probability cannot be infinite; total probability is unity. This is to assert that the wavefunction is a member of a complete orthogonal set of functions, L^2.

Any function that is continuous, differentiable, and square integrable can be a wavefunction.

We assume that the wavefunction is the de Broglie wave:

$$\psi = Ae^{-i\left(\frac{Et-px}{\hbar}\right)}$$

$$= \begin{bmatrix} A & 0 \\ 0 & A \end{bmatrix} \begin{bmatrix} \cos\left(\dfrac{Et-px}{\hbar}\right) & -\sin\left(\dfrac{Et-px}{\hbar}\right) \\ \sin\left(\dfrac{Et-px}{\hbar}\right) & \cos\left(\dfrac{Et-px}{\hbar}\right) \end{bmatrix} \qquad (21.1)$$

Subject to the above assumptions, we try a few possible wave equations. This is not logical deduction; it is guesswork. We consider the wave equation:

$$\frac{\partial^2 \psi}{\partial x^2} = \frac{1}{v^2} \frac{\partial^2 \psi}{\partial t^2} \tag{21.2}$$

Substituting the de Broglie wave into this leads to:

$$E^2 = p^2 v^2 \tag{21.3}$$

This is the relativistic energy expression for a massless particle (moving at the speed of light). Since we are concerned with particles of non-zero mass, we must reject the above wave equation, (21.2).

21.1 THE TIME DEPENDENT SCHRÖDINGER EQUATION

The relativistic energy expression contains the mass-energy term and the classical kinetic energy term and the higher powers of the classical kinetic energy term. If we extract from that relativistic energy expression only the classical kinetic energy term, we have:

$$Kinetic\ Energy = \frac{p^2}{2m} \tag{21.4}$$

We are constructing a non-relativistic theory, and so we seek a wave equation that is consistent with this non-relativistic energy expression. We begin with the de Broglie wave equation and we differentiate it with respect to both time and space:

$$\psi = Ae^{-i\left(\frac{Et-px}{\hbar}\right)}$$

$$\frac{\partial \psi}{\partial t} = -iE\frac{1}{\hbar}Ae^{-i\left(\frac{Et-px}{\hbar}\right)} \quad : \quad \frac{\partial^2 \psi}{\partial x^2} = -p^2\frac{1}{\hbar^2}Ae^{-i\left(\frac{Et-px}{\hbar}\right)} \tag{21.5}$$

Rearranging:

$$i\frac{\partial \psi}{\partial t}\hbar\frac{1}{A}e^{i\left(\frac{Et-px}{\hbar}\right)} = E \quad : \quad -\hbar^2\frac{1}{A}\frac{\partial^2 \psi}{\partial x^2}e^{i\left(\frac{Et-px}{\hbar}\right)} = p^2 \tag{21.6}$$

Putting these expressions into the kinetic energy expression, (21.4), gives:

$$i\frac{\partial \psi}{\partial t} = -\frac{\hbar}{2m}\frac{\partial^2 \psi}{\partial x^2} \tag{21.7}$$

This equation is consistent with the non-relativistic energy expression. $i = \sqrt{-1}$, ψ is the wavefunction associated with the particle, and m is the mass of the particle. This equation is known as the zero potential time dependent Schrödinger equation, TDSE.

The time dependent Schrödinger equation describes the time evolution of the wavefunction – what happens to the particle of mass m associated with the wavefunction. The TDSE appropriate to a particular physical system contains all the information about results of measurements (observations) of that particular physical system.

The TDSE is not the only wave equation that is consistent with the classical energy expression, but it is an equation that is linear. The fact that the TDSE is linear allows us to add different solutions of it together to form more solutions of it. This means that we can superimpose solutions to describe the interference that we see in electron diffraction experiments. Linearity is a very important property of the TDSE.

Because the TDSE is linear, it has basis solutions which form a linear space that contains all its solutions. There would be little point in assuming that wavefunctions form the linear space L^2 if we did not make the wave equation linear. Since probability is finite, if we allow the probability interpretation of the modulus of the wavefunction, we have to assume that wavefunctions form the linear space L^2.

As it stands, the time dependent Schrödinger equation, TDSE, is valid for a free particle – that is a particle moving through a zero (uniform) potential. We modify the above TDSE to describe a particle in a potential by adding the potential energy. In a non-uniform (non-zero) potential, $V(t, x)$, the TDSE is:

$$i\frac{\partial \psi}{\partial t} = -\frac{\hbar}{2m}\frac{\partial^2 \psi}{\partial x^2} + V(t,x)\psi \tag{21.8}$$

$V(t, x)$ is the potential energy of a particle with mass m at the point in space, x, at the time t. This equation is also linear.

Because the TDSE is of first order with respect to time, the state of the system it describes at some initial time, t_0, determines the behavior of that system for all future time; the TDSE is deterministic.

Since there is a $i = \sqrt{-1}$ on one side of the equation, there must be a $i = \sqrt{-1}$ on the other side of the equation. This means that ψ is a complex function:

$$\psi = \begin{bmatrix} f(t,x) & g(t,x) \\ -g(t,x) & f(t,x) \end{bmatrix} = f(t,x) + i.g(t,x) \tag{21.9}$$

Aside: In matrix notation, the TDSE is:

$$\begin{bmatrix} 0 & 1 \\ -1 & 0 \end{bmatrix} \begin{bmatrix} \dfrac{\partial f}{\partial t} & \dfrac{\partial g}{\partial t} \\ -\dfrac{\partial g}{\partial t} & \dfrac{\partial f}{\partial t} \end{bmatrix} = \begin{bmatrix} -\dfrac{\hbar}{2m} & 0 \\ 0 & -\dfrac{\hbar}{2m} \end{bmatrix} \begin{bmatrix} \dfrac{\partial^2 f}{\partial x^2} & \dfrac{\partial^2 g}{\partial x^2} \\ -\dfrac{\partial^2 g}{\partial x^2} & \dfrac{\partial^2 f}{\partial x^2} \end{bmatrix}$$

$$+ \begin{bmatrix} V & 0 \\ 0 & V \end{bmatrix} \begin{bmatrix} f & g \\ -g & f \end{bmatrix} \tag{21.10}$$

$$\begin{bmatrix} -\dfrac{\partial g}{\partial t} & \dfrac{\partial f}{\partial t} \\ -\dfrac{\partial f}{\partial t} & -\dfrac{\partial g}{\partial t} \end{bmatrix} = \begin{bmatrix} -\dfrac{\hbar}{2m}\dfrac{\partial^2 f}{\partial x^2} & -\dfrac{\hbar}{2m}\dfrac{\partial^2 g}{\partial x^2} \\ \dfrac{\hbar}{2m}\dfrac{\partial^2 g}{\partial x^2} & -\dfrac{\hbar}{2m}\dfrac{\partial^2 f}{\partial x^2} \end{bmatrix}$$

$$+ \begin{bmatrix} Vf & Vg \\ -Vg & Vf \end{bmatrix} \tag{21.11}$$

Which we can write as:

$$\frac{\partial g}{\partial t} = \frac{\hbar}{2m}\frac{\partial^2 f}{\partial x^2} - Vf \tag{21.12}$$

$$\frac{\partial f}{\partial t} = -\frac{\hbar}{2m}\frac{\partial^2 g}{\partial x^2} + Vg$$

We have above made an attempt to derive the Schrödinger equation. Perhaps we should not have so spent our time. It is more direct, and perhaps more honest to simple admit that we postulate the TDSE and we make no attempt to justify it other than "it works," and it works very well, for non-relativistic particles. Having postulated the Schrödinger equation, we could then deduce de Broglie's wave equation from it.

We can think of the TDSE as the quantum mechanical equation of motion. It controls (deterministically) how a physical system evolves in time.

21.2 THE TDSE IN THREE DIMENSIONS

In three dimensions, the TDSE is:

$$i\frac{\partial \psi(t,r)}{\partial t} = -\frac{\hbar}{2m}\nabla^2\psi(t,r) + V(t,r)\psi(t,r) \qquad (21.13)$$

Where $\nabla^2\psi = \dfrac{\partial^2\psi}{\partial x^2} + \dfrac{\partial^2\psi}{\partial y^2} + \dfrac{\partial^2\psi}{\partial z^2}$

21.3 THE TIME INDEPENDENT SCHRÖDINGER EQUATION, TISE

There are many physical systems in which the potential, $V(t, x)$, depends on time. A fluctuating electromagnetic field associated with electromagnetic radiation is such a potential that varies with time. There are many physical systems in which the potential, $V(x)$, does not depend on time but depends upon only position[3] in space. An electron orbiting a stable atom isolated from any electromagnetic radiation is in such a potential that does not vary with time. If the potential, $V(x)$, is independent of time, then it is always possible to find separable solutions of the time dependent Schrödinger equation of the form:

$$\psi(r,t) = u(r)T(t) \qquad (21.14)$$

Where $u(r)$ is a function of only spatial position and $T(t)$ is a function of only time. The reader is reminded of the Newtonian wave equa-

---

[3.] It is a normal notational quirk of quantum mechanics, and of QFT, to write a function of three, or more, spatial co-ordinates with a single spatial co-ordinate.

192 • Quantum Mechanics

tion. The function $\psi(r, t) = u(r)T(t)$ is called a wavefunction. Confusingly, the function $u(r)$ is also called a wavefunction.

The solution $\psi(r, t) = u(r)T(t)$ is a standing wave – thus we see an electron in orbit around a nucleus to be a standing wave. Substituting the solution (21.14) into the time dependent Schrödinger equation leads to:

$$-\frac{\hbar}{i}\frac{1}{T(t)}\frac{\partial T(t)}{\partial t} = \frac{1}{u(r)}\left(-\frac{\hbar^2}{2m}\frac{\partial^2 u(r)}{\partial x^2} + V(r)u(r)\right) \quad (21.15)$$

This equation is a function of only time on the left-hand side and of only position on the right-hand side. Position and time are independent of each other, and we must therefore have that both sides are equal to a constant, K. This leads to:

$$-\frac{\hbar}{i}\frac{\partial T(t)}{\partial t} = KT(t)$$

$$\left(-\frac{\hbar^2}{2m}\frac{\partial^2 u(r)}{\partial x^2} + V(r)u(r)\right) = Ku(r) \quad (21.16)$$

These are eigenvalue equations. The constant (eigenvalue), K, is just the energy.

The second of these, (21.16), is called the time independent Schrödinger equation, TISE, - which is different from the TDSE because everything in it is independent of time.

We did not start off by saying that we were interested in only eigenvalue equations. We started off looking at all types of solutions to the TDSE and were driven to the TISE by seeking separable solutions. That the TISE is an eigenvalue equation is not by our choice that it should be so. It just turns out that solutions of the TISE for time independent potentials are eigenvalue equations. This is how eigenvalues get into physics, not because we put them there but because the basis solutions of linear differential equations are separable and thus equal to a constant, K.

Rewriting the above (21.16) gives:

$$-\frac{\hbar}{i}\frac{\partial}{\partial t}(T(t)) = E_n T(t) \quad (21.17)$$

And:

$$\left(-\frac{\hbar^2}{2m}\frac{\partial^2}{\partial x^2}+V(r)\right)\psi(r)=E_n\psi(r) \qquad (21.18)$$

The time independent Schrödinger equation is:

$$-\frac{\hbar^2}{2m}\frac{\partial^2\psi(r)}{\partial x^2}+V(r)\psi(r)=E_n\psi(r) \qquad (21.19)$$

The first of the above pair of equations, (21.17), might be called the position independent Schrödinger equation, but it isn't; this equation is easily solved to give:

$$T(t)=e^{-i\frac{Et}{\hbar}} \qquad (21.20)$$

We therefore have that the solution (wavefunction) of the time dependent Schrödinger equation is:

$$\psi(r,t)=u(r)T(t)=u(r)e^{-i\frac{Et}{\hbar}} \qquad (21.21)$$

We reiterate, if the potential, $V(x)$, is independent of time, the solution of the TDSE is of the form $\psi(x,t)=\Phi(x)e^{-i\frac{Et}{\hbar}}$ such that $\Phi(x)$ is a solution of the TISE.

If ψ is a solution of the TDSE, then ψ^* is also a solution of the TDSE equation. Also, if ψ is a solution of the TISE, then ψ^* is also a solution of the TISE equation.

Aside: The 3-dimensional TISE in polar co-ordinates with a Coulomb potential is:

$$\left(-\frac{\hbar^2}{2m}\nabla^2-\frac{e^2}{r}\right).u_{E_n}(r,\theta,\phi)=E_n u_{E_n}(r,\theta,\phi) \qquad (21.22)$$

We will meet this when we look at the hydrogen atom towards the end of this book. That this equation correctly predicts the energy levels of electrons within the hydrogen atom is a strong verification of the theory of quantum mechanics in spite of all the theory's weird operator type mathematics.

21.4 STATIONARY STATES

Stationary states are states in which the probability (intensity of the wave) is independent of time. This means that, for a physical system in a stationary state, the probability of a particular outcome of measuring a dynamical variable of that physical system does not change with time – the physical system does not change with time.

We can think of the standing wave solutions of the Newtonian ideal string equation as stationary states. The stationary states and the basis solutions are the same things. Well, one is a physical state and the other is a solution to an equation, but the idea is clear.

Stationary states are solutions of the TDSE when the potential, $V(x)$, is independent of time. Stationary states of the TDSE are of the form:

$$\Psi(x,t) = \phi(x)e^{-i\frac{Et}{\hbar}} \tag{21.23}$$

They are called stationary states because the scalar field that is the probability density (wave intensity) does not vary with time. With each solution of the TDSE, there is an associated probability density defined by $P = |\psi|^2 = \psi^*\psi$. If the solution is of the form $\psi(x,t) = \Phi(x)e^{-i\frac{Et}{\hbar}}$, then the associated probability density is:

$$P = \psi^*\psi$$
$$= \Phi^*(x)e^{i\frac{Et}{\hbar}}\Phi(x)e^{-i\frac{Et}{\hbar}} \tag{21.24}$$
$$= \Phi^*(x)\Phi(x)$$

which is independent of time. An example of such a solution with zero potential would be:

$$\psi = A\sin(kx)e^{-i\frac{Et}{\hbar}} \tag{21.25}$$

Or:

$$\psi^* = A\sin(kx)e^{i\frac{Et}{\hbar}} \tag{21.26}$$

If we feed this solution into the TDSE we get:

$$i\hbar \frac{\partial\left(A\sin(kx)e^{-i\frac{Et}{\hbar}}\right)}{\partial t} = -\frac{\hbar^2}{2m} \frac{\partial^2\left(A\sin(kx)e^{-i\frac{Et}{\hbar}}\right)}{\partial x^2}$$

$$i\hbar\left(A\sin(kx)\right)\left(-i\frac{E}{\hbar}e^{-i\frac{Et}{\hbar}}\right) = -\frac{\hbar^2}{2m}\left(-Ak^2\sin(kx)\right)e^{-i\frac{Et}{\hbar}} \quad (21.27)$$

$$Ee^{-i\frac{Et}{\hbar}}A\sin(kx) = \frac{\hbar^2 k^2}{2m}e^{-i\frac{Et}{\hbar}}A\sin(kx)$$

This is a solution if $E = \dfrac{\hbar^2 k^2}{2m}$: $k \in \mathbb{R}$.

Doing it a different way, let the energy operator, the Hamiltonian, act upon this solution:

$$-\frac{\hbar^2}{2m}\frac{\partial^2\left(e^{-i\frac{Et}{\hbar}}A\sin(kx)\right)}{\partial x^2} = \frac{\hbar^2 k^2}{2m}e^{-i\frac{Et}{\hbar}}A\sin(kx) \quad (21.28)$$

And we find that the solution is an eigenfunction of the Hamiltonian (the energy operator) with eigenvalues $\dfrac{\hbar^2 k^2}{2m}$.

The modulus squared of this solution is:

$$|\psi|^2 = \psi^* \psi$$

$$= \left(A\sin(kx)e^{i\frac{Et}{\hbar}}\right)\left(A\sin(kx)e^{-i\frac{Et}{\hbar}}\right) \quad (21.29)$$

$$= A^2\sin^2(kx)e^{i\frac{Et}{\hbar}}e^{-i\frac{Et}{\hbar}}$$

$$= A^2\sin^2(kx)$$

Notice how taking the conjugate has ridded us of the time part of this solution. The modulus squared is now independent of time – it is a standing wave. The modulus squared (wave intensity) evaluated at point, x, is proportional to the probability that the particle will be found at the point x at any time. This probability (wave intensity) does not vary with time.

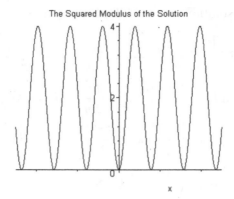

The Squared Modulus of the Solution

We see that, at all times, there are points associated with zero probability of the particle being at those points – the particle can never be there. There are also points where there is a lot of probability of the particle being at those points.

21.5 GROUND STATE

Of the above solutions, there is one solution, with $k = 1$, with eigenvalue $E_1 = \dfrac{\hbar^2}{2m}$. This is the lowest possible value of energy. The lowest energy solution is known as the ground state. In our example above, the other energies are multiples of this state but this is very exceptional. It is not common to find that energies are multiples of the ground state. The $k = 0$ solution is $\psi = A\sin(0)e^{-i\frac{Et}{\hbar}} = 0$. Zero energy is the same as non-existence, and so we do not find particles with zero energy.

21.6 NORMALIZATION

The interpretation of $|\psi|^2$ as the probability leads to the need to normalize the sum of all the $|\psi|^2$ to unity because, by convention, total probability is unity. We have above that the probability of our particle being at a point x is given by:

$$\text{Probability} = |\psi|^2 = A^2 \sin^2(kx) \qquad (21.30)$$

If we add the probabilities of the particle being at each point in the interval $[a, b]$, then we will get the probability of the particle being between those two points. We add by integrating:

$$\text{Probability}_{(a,b)} = \int_a^b dx \, |\psi|^2$$

$$= A^2 \int_a^b dx \, \sin^2(kx) \qquad (21.31)$$

$$= \frac{A^2}{k} \big(\cos(ak)\sin(ak) - \cos(bk)\sin(bk)\big) + A^2(b-a)$$

Since total probability must be unity, we must have:

$$\text{Total Probability} = \int_{-\infty}^{\infty} dx \, |\psi|^2 = 1$$

$$= A^2 \int_{-\infty}^{\infty} dx \, \sin^2(kx) = 1 \qquad (21.32)$$

That this integral is undefined between $(-\infty, \infty)$ reflects the fact that this is not a solution which exists in an unbounded system. The integral is defined between any two points, and this reflects the fact that this is a solution within a bound system. We achieve the equality with unity by adjusting the value of the amplitude, A.

21.7 THE NORMALIZATION CONDITION

The above requirement that total probability equals unity is the same thing as requiring that the particle is somewhere in the universe. This requirement is referred to as the normalization condition:

$$\int_{-\infty}^{\infty} dx \, |\psi|^2 = 1 \qquad (21.33)$$

As above, it is achieved by adjusting the amplitude of the wavefunction – that is multiplying it by a constant.

Technically, every solution of a linear equation can be multiplied by a number (also called a scalar - an element of a division algebra – it could be a complex number) and remain a solution of that equation. None of the physical properties described by that solution are changed by such scalar multiplication. Normalization is just scaling the solutions.

To reiterate: we normalize the wavefunction $\psi(t, x)$ by solving:

$$A^2 \int_{-\infty}^{\infty} dx \ |\psi|^2 = A^2 \int_{-\infty}^{\infty} dx \ \psi^* \psi = 1 \qquad (21.34)$$

For $A \in \mathbb{R}$. The normalized wavefunction is then $A. \ \psi(t, x)$

If the particle is confined to a region $0 < x < L$, then we integrate between the extremities of this region, and the normalizing condition is:

$$A^2 \int_0^L dx \ |\psi|^2 = A^2 \int_0^L dx \ \psi^* \psi = 1 \qquad (21.35)$$

WORKED EXAMPLE

Normalize the momentum eigenfunction $u_p(x) = Ce^{i\frac{p}{\hbar}x}$ for a particle in the interval $0 < x < L$. We have the normalization condition:

$$\int_0^L dx \ |\psi|^2 = 1$$

$$|C|^2 \int_0^L dx \ e^{-i\frac{p}{\hbar}x} e^{i\frac{p}{\hbar}x} = 1$$

$$|C|^2 \int_0^L dx \ 1 = 1 \qquad (21.36)$$

$$|C|^2 [x]_0^L = 1$$

$$C = \frac{1}{L^{\frac{1}{2}}}$$

The normalized momentum eigenfunction is:

$$u_p(x) = \frac{1}{L^{\frac{1}{2}}} e^{i\frac{p}{\hbar}x} \qquad (21.37)$$

Note that the relative probability of finding a particle at two different places does not involve the normalization constant as the normalization constant will cancel when we calculate the ratio of the two probabilities.

21.8 DETERMINISM AND THE SCHRÖDINGER EQUATION

The Schrödinger equation is a deterministic equation in that, given the initial conditions of a physical system it describes how that system evolves in time. That description is a definite mathematical expression that has no concept of probability within it. However, the physical state is described by a definite complex expression, and we are unable to observe the phase of the expression. So, although the mathematics is deterministic, our interpretation (observation) of that mathematics is unable to see the whole of the predetermined expression in physical reality. Probability comes into quantum mechanics because the Schrödinger equation has solutions that are complex. This means that the Schrödinger equation cannot predict the fate of a single electron but can statistically predict the combined fate of millions of electrons.

Aside: A die does not remember what number was rolled on a previous casting of it. So, how does it know that when rolled six million times it should produce approximately a million sixes? Is this conservation of sixes similar to conservation of energy? Dice do not communicate with each other. So, if six million of them are rolled together, how do they know to produce approximately a million sixes? So it is with the collapse of the wavefunction. Ah! The mysteries of the universe are more numerous than the stars that bedeck the firmament.

21.9 THE STRUCTURE OF THE WAVEFUNCTION

Remember, because the time dependent Schrödinger equation, TDSE, is linear, any sum of solutions of it is also a solution of it.

This sum of solutions is weighted with coefficients that multiply each separate solution. We consider only separable solutions (separable solutions are the basis solutions) of the TDSE – those of the form $\psi_n e^{\frac{-E_n t}{\hbar}}$. Thus, every solution of the TDSE of the form $\Psi e^{\frac{-E}{\hbar}t}$, that is every wavefunction of the TDSE of the form $\Psi e^{\frac{-E}{\hbar}t}$, can be written as:

$$\Psi e^{\frac{-E}{\hbar}t} = \sum_{n=0...} c_n \psi_n e^{\frac{-E_n t}{\hbar}} \qquad (21.38)$$

At $t = 0$, this is a linear sum of stationary solutions:

$$\Psi_{t=0} = \sum_{n=0...} c_n \psi_n \qquad (21.39)$$

The stationary states are called basis states because, at $t = 0$, any separable wavefunction can be written as a sum of them. As time passes, the solutions of the TDSE evolve as determined by the $e^{\frac{-E_n t}{\hbar}}$ factor.

21.10 GLOBAL PHASE AND LOCAL PHASE

Take a wavefunction:

$$\Psi(x) = c_1 \psi_1 + c_2 \psi_2 + c_3 \psi_3 + \qquad (21.40)$$

If this is wholly multiplied by a phase, there is no observable difference between the original wavefunction and the new wavefunction:

$$\Psi = e^{i\theta} c_1 \psi_1 + e^{i\theta} c_2 \psi_2 + e^{i\theta} c_3 \psi_3 +$$
$$\Psi^\circ \Psi = e^{-i\theta} \Psi(x) e^{i\theta} \Psi(x) \qquad (21.41)$$

We say that the wavefunctions are related by a global phase. However, if only one (or some) of the basis wavefunctions are multiplied by a phase, then the new wavefunction is not observationally equivalent to the old wavefunction.

$$\Psi \neq \Psi' = c_1 \psi_1 + e^{i\theta} c_2 \psi_2 + c_3 \psi_3 + \qquad (21.42)$$

We refer to this a local phase change, and we are able to observe the difference.

21.11 MOVING PARTICLES

The sums of stationary solutions are not stationary. Particles that are not bound within a potential (free particles) are described not by stationary solutions but by (Fourier type) sums of stationary solutions. Consider a solution that is a sum of just two stationary solutions:

$$\Psi(t,x) = A\psi_1 e^{-\frac{E_1}{\hbar}t} + B\psi_2 e^{-\frac{E_2}{\hbar}t} \quad : \quad E_1 \neq E_2 \qquad (21.43)$$

The probability density of this solution is:

$$|\Psi|^2 = \Psi^{\circ}\Psi$$

$$= |A|^2 |\psi_1|^2 + |B|^2 |\psi_2|^2 \qquad (21.44)$$

$$+ AB^{\circ}\psi_1\psi_2^{\circ} e^{i\frac{(E_2-E_1)t}{\hbar}} + A^{\circ}B\psi_1^{\circ}\psi_2 e^{i\frac{(E_1-E_2)t}{\hbar}}$$

We see that, unless $E_1 = E_2$, the probability density varies with time. This means that the physical system (say, a particle) is moving.

To form a particle like wavefunction (a wave packet), we need to add a lot more than just two basis solutions; we normally take it that an infinite number of basis solutions are added to produce a particle. The group velocity of the wave packet is taken to be the velocity of the particle.

EXERCISES

1. Given $\Psi(t,x) = A(4x^3 + 2x)e^{-i\frac{Et}{\hbar}}$, and assuming that this is a solution of the TDSE, find the potential, $V(x)$?

2. a. For the wavefunction $\Psi(x) = A\cos\left(\frac{2\pi}{a}x\right)$, find the normalization constant A?

b. What is the probability that the particle described by the wavefunction in 2a will be found in the interval $\left[\frac{a}{3}, \frac{a}{2}\right]$?

3. Given $\phi(x) = A\dfrac{1+ix}{1+ix^2}$, normalize $\phi(x)$, and find the probability that the particle it describes will be found in the interval $x = [0, 1]$. You will need: $\displaystyle\int_{-\infty}^{\infty} dx\, \frac{1+x^2}{1+x^4} = \sqrt{2}\pi$.

22

SOLVING THE TISE IN SEVERAL POTENTIALS

Over the next few chapters, we will solve the time independent Schrödinger equation, TISE, in various potentials. When we choose to use the Schrödinger equation, we are choosing to accept that a particle is described by a wave equation – we are choosing that a particle is really a wave. It is not surprising, having chosen that a particle is a wave, to discover that the wave equation that describes that particle has solutions that are waves, and so the particle is a wave.

We are going to look at physical systems in which the potential is not varying over time. Because the TISE is "only part" of the full Schrödinger equation, the TDSE, we can use it only when the potential does not vary with time – it is the "part" that is time independent. An example of such a stationary state would be the hydrogen atom (or any other atom) or a free particle.

Aside: Another way of saying that the potential is not varying with time is to say that the Hamiltonian (energy operator) is not varying in time, and the reader might often hear the justification for using the TISE phrased in terms of the Hamiltonian being independent of time.

When we have found solutions, ψ, of the TISE, we can turn them into solutions of the "whole" time dependent Schrödinger equation,

TDSE, by multiplying them by $e^{-i\frac{Et}{\hbar}}$. When we use the TISE, we are assuming that the solutions of the TDSE are separable because the TISE was deduced upon that assumption; we deal with only such solutions.

We will need the following:

$$p = \frac{h}{\lambda} = \frac{h}{2\pi}\frac{2\pi}{\lambda} = \hbar k \qquad (22.1)$$

Where p is the momentum of a particle and k is the wave number of the particle. From this, using the classical kinetic energy expression $E = \frac{p^2}{2m}$, we get:

$$E = \frac{p^2}{2m} = \frac{\hbar^2 k^2}{2m} \qquad (22.2)$$

THE FREE PARTICLE SOLUTION OF THE TISE

For a particle in potential V, the TISE is the energy eigenvalue equation:

$$\widehat{H}\psi_n = E_n\psi_n$$

$$-\frac{\hbar^2}{2m}\frac{\partial^2\psi_n}{\partial x^2} + V\psi_n = E_n\psi_n \qquad (23.1)$$

Where E_n is the energy of the particle, m is the mass of the particle and ψ_n is an eigenfunction that describes the state of the particle. $\widehat{H} = -\dfrac{\hbar^2}{2m}\dfrac{\partial^2}{\partial x^2} + V$ is the Hamiltonian, otherwise known as the energy operator. For a free particle, the potential is zero (or uniform if you prefer), and the TISE becomes the eigenvalue equation:

$$-\frac{\hbar^2}{2m}\frac{\partial^2\psi_n}{\partial x^2} = E_n\psi_n \qquad (23.2)$$

We seek to find the eigenfunctions, ψ_n, and the associated eigenvalues, E_n. Such eigenfunctions will be the basis of a linear space because the TISE is a linear differential equation. The wavefunction of the particle is the weighted sum of all the eigenfunctions, but we will never observe the wavefunction. What we will observe is one of the eigenfunctions with its single eigenvalue. When we seek a list of those possible eigenfunctions and their eigenvalues, we seek a list of possible outcomes of an observation to measure the energy of the particle – so we can predict what we might get.

We have from (23.2):

$$\frac{\partial^2 \psi_n}{\partial x^2} + \frac{2m}{\hbar^2} E_n \psi_n = 0 \tag{23.3}$$

Using (23.2) leads to:

$$\frac{\partial^2 \psi_n}{\partial x^2} + k_n^2 \psi_n = 0 \tag{23.4}$$

This equation has real and complex solutions, but the essence is caught by the two solutions:

$$\psi_1 = Ae^{ikx} \qquad : \qquad \psi_2 = Ae^{-ikx}$$
$$\psi_1 = Ae^{i\frac{\sqrt{2mE}}{\hbar}x} \qquad : \qquad \psi_2 = Ae^{-i\frac{\sqrt{2mE}}{\hbar}x} \tag{23.5}$$

Both rotations; one is clockwise, and the other is counterclockwise.

Aside: In matrix notation, the two eigenfunctions of a free particle are:

$$\psi_1 = \begin{bmatrix} A & 0 \\ 0 & A \end{bmatrix} \begin{bmatrix} \cos(kx) & \sin(kx) \\ -\sin(kx) & \cos(kx) \end{bmatrix}$$
$$\psi_2 = \begin{bmatrix} A & 0 \\ 0 & A \end{bmatrix} \begin{bmatrix} \cos(kx) & -\sin(kx) \\ \sin(kx) & \cos(kx) \end{bmatrix} \tag{23.6}$$

The above two solutions are solutions for all values of energy. There are no boundary conditions to satisfy because we are dealing with a free particle. We therefore have an infinite number of eigenfunctions and a corresponding infinite number of eigenvalues. The set of eigenvalues spans the whole of the real numbers. The free particle can have any energy.

23.1 DEGENERACY

We actually have two eigenfunctions for each eigenvalue. For any value of $k = \frac{\sqrt{2mE}}{\hbar}$, we have both clockwise rotation and counterclockwise rotation. This situation in which two or more eigenfunctions (solutions of an eigenvalue equation) have the same eigenvalue is called degeneracy.

23.2 FREE PARTICLE MOMENTUM EIGENFUNCTIONS

We have two wavefunctions (energy eigenfunctions) that solve the TISE for a free particle. For a free particle, these two solutions are also eigenfunctions of the momentum operator, $\hat{p} = -i\hbar\dfrac{\partial}{\partial x}$. We have:

$$\hat{p}_x(\psi_1) = -i\hbar\frac{\partial}{\partial x}(Ae^{ikx}) = -i\hbar Aike^{ikx} = \hbar kAe^{ikx} = \hbar k\psi_1 \qquad (23.7)$$

We see that the momentum eigenvalues are $\hbar k$, which, again, can take any value, and so a free particle can have any value of momentum. Isaac Newton knew that! The other wavefunction has eigenvalues $-\hbar k$. The plus and minus signs correspond to the direction of travel, to the right or to the left, of the particle.

We have seen that the energy eigenfunctions of a free particle, $\{\psi_1, \psi_2\}$, are also its momentum eigenfunctions. This implies that, for a free particle, the energy operator (the Hamiltonian) and the momentum operator commute with each other and it is possible to simultaneously know both the momentum and the energy of a free particle.

23.3 THE FREE PARTICLE SOLUTIONS OF THE TDSE ARE NOT SQUARE INTEGRABLE

There is a problem that we have glossed over. When we use the TISE, we assume that the solutions to the TDSE are separable. The TISE was derived on the basis of this assumption. Thus, the solutions we have are separable solutions. However, the separable solutions of the free particle (zero potential) TDSE are not square integrable and therefore cannot be wavefunctions. The reader should recall that the squared modulus of the wavefunction is proportional to the probability density and the sum of all probability densities at every point in space must be finite. We have:

$$\int_{-\infty}^{\infty} dx \ |\psi_1|^2 = \int_{-\infty}^{\infty} dx \ Ae^{-i\frac{\sqrt{2mE}}{\hbar}x} Ae^{i\frac{\sqrt{2mE}}{\hbar}x} = A^2 \int_{-\infty}^{\infty} dx \ 1 = \infty \quad (23.8)$$

There are no separable solutions of the TDSE that describe a free particle. There are non-separable solutions of the TDSE that describe a free particle. The linearity of the TDSE allows such solutions to be constructed by adding the separable solutions. We will put that slightly differently. The solutions of the TDSE for a free particle (zero potential) are not basis solutions of the linear space of solutions. The solutions of the TDSE for a free particle are sums of basis solutions of the linear space of solutions. Separable solutions and basis solutions are the same things.

The separable solutions of the TDSE are the solutions of the TISE multiplied by $e^{-i\frac{Et}{\hbar}}$. The sum of such solutions is a solution of the TDSE of the form:

$$\sum e^{-i\frac{Et}{\hbar}}\psi_i = \sum e^{-i\frac{E_n t}{\hbar}} A_n e^{i\frac{\sqrt{2mE_n}}{\hbar}x} = \sum A_n e^{i\frac{1}{\hbar}\left(-E_n t + \sqrt{2mE_n}x\right)} \quad (23.9)$$

This expression might look posh, but we have seen it in a simpler form before. It is no more than a linear sum:

$$\Psi = e^{-i\frac{Et}{\hbar}} (c_1\psi_1 + c_2\psi_2 + c_3\psi_3 + ...) \quad (23.10)$$

In matrix form, this is:

$$\sum \begin{bmatrix} A_n & 0 \\ 0 & A_n \end{bmatrix} \begin{bmatrix} \cos\left(\frac{-E_n t + \sqrt{2mE_n}}{\hbar}x\right) & \sin\left(\frac{-E_n t + \sqrt{2mE_n}}{\hbar}x\right) \\ -\sin\left(\frac{-E_n t + \sqrt{2mE_n}}{\hbar}x\right) & \cos\left(\frac{-E_n t + \sqrt{2mE_n}}{\hbar}x\right) \end{bmatrix} \quad (23.11)$$

This is just a sum of de Broglie waves. Basically, we are just adding Euclidean trigonometric functions in a Fourier series to produce a wave packet. The wave packet has particle-like properties, and that is why we think it to be a particle.

To put it slightly differently, the unbound solutions (free particle solutions) are not normalizable and therefore cannot represent a state of the system. We have:

$$|\Psi|^2 = \left|Ae^{i(kx-\omega t)}\right|^2 = |A|^2 \qquad (23.12)$$

But such solutions can be added to form a wave packet that is both localized and normalizable. The sum of such solutions is a superposition of solutions. A superposition of (non-degenerate) stationary states is not a stationary state, and it does not represent an unchanging physical system; it represents a moving particle.

Aside: There is a thing of considerable philosophical weight here. Any old linear sum of basis solutions will not form a wave packet. To form a wave packet, we need exactly the correct coefficients, c_i, and only some (a particular set) of basis solutions. How does nature get only the required basis solutions and use exactly the correct coefficients to form a wave packet?

23.4 SUMMARY OF FREE PARTICLE SOLUTIONS OF THE TDSE

We are interested in only solutions of the "whole" Schrödinger equation, the TDSE. There are separable solutions to the "whole" TDSE for free particles (particles in zero potential), but these solutions are not wavefunctions because they are not square integrable (not normalizable). We are interested in only the solutions that are wavefunctions – that is elements of the square integrable functions, L^2.

There are non-separable solutions of the TDSE for free particles that are elements of the square integrable functions, L^2. These solutions are sums of separable solutions. We take it that these sums are such that they form a wave packet.

Free particles are wave packets that are linear sums of separable solutions (basis solutions) of the free particle, $V = 0$, TDSE.

24

THE INFINITE SQUARE WELL SOLUTION OF THE TDSE

We would like to solve the TDSE for an electron in the Coulomb potential of an atomic nucleus. This would give us the energy levels of the orbiting electrons. The differences between those energy levels will be the energies of the electromagnetic spectrum of the atom. We will deal with such a case when we look at the hydrogen atom. Meanwhile, we approximate the Coulomb potential of the atomic nucleus with an infinite square well potential. We will get qualitative results, but the numbers will be inaccurate.

We want to know the energies (energy levels) of a 1-dimensional square box with infinitely high walls and a constant zero potential between these walls.

A Square Well Potential

The potential is given by:

$$V(x) = 0 \quad : \quad 0 \le x \le a$$
$$V(x) = \infty \quad : \quad x > a \ \& \ x < 0 \quad\quad (24.1)$$

The height of the potential barrier is such that no particle (think rubber ball) can get out of the well. Whatever the energy of the particle, it will "bounce non-elastically" from the "walls" of the potential well. (There is no quantum tunneling through an infinite potential.) The potential is independent of time, and so we seek solutions of the TISE. Since, this 1-dimensional potential is defined in three parts, there are three time independent Schrödinger equations – one for each part of the potential – but we have no interest in the parts of the potential that are outside of the well, $\{x < 0, x > a\}$. The time independent Schrödinger equation, TISE, inside the well is:

$$\left(-\frac{\hbar^2}{2m}\frac{\partial^2}{\partial x^2} + V(x)\right)\left(u_E(x)\right) = Eu_E(x)$$

$$\left(-\frac{\hbar^2}{2m}\frac{\partial^2}{\partial x^2} + 0\right)\left(u_E(x)\right) = Eu_E(x) \quad\quad (24.2)$$

$$\frac{\partial^2 u_E}{\partial x^2} + \frac{2mE}{\hbar^2}u_E = 0$$

This is the same TISE as for a free particle (because the potential in both cases was zero). However, this is not a free particle because there are boundary conditions.

We presume that the energy eigenfunctions, u_E, will be zero at $x = 0$ & $x = a$, which is at the walls where the potential is infinite just as the kinetic energy of a bouncing rubber ball would be zero at the wall. We impose this assumption upon the solutions as a boundary condition on the system. Of all possible solutions, we choose to use the most general solution:

$$u = A\cos\left(\frac{\sqrt{2mE}}{\hbar}x\right) + B\sin\left(\frac{\sqrt{2mE}}{\hbar}x\right) \quad\quad (24.3)$$

As we have seen above for the free particle, without the boundary condition, the energy eigenvalues (energy levels), $\{E_{2n}, E_{2n+1}\}$ could be anything. The energy eigenvalues are spread across the whole of the real numbers. However, with the infinite square well potential, we need to satisfy the boundary condition that $u_E = 0: x = 0$. We have:

$$A\cos\left(\frac{\sqrt{2mE}}{\hbar}0\right) + B\sin\left(\frac{\sqrt{2mE}}{\hbar}0\right) = 0 \quad \Rightarrow \quad A = 0 \qquad (24.4)$$

Thus, we have reduced the solution to:

$$u = B\sin\left(\frac{\sqrt{2mE}}{\hbar}x\right) \qquad (24.5)$$

We now impose the other boundary condition $u_E = 0: x = a$:

$$B\sin\left(\frac{\sqrt{2mE}}{\hbar}a\right) = 0 \quad \Rightarrow \quad \frac{\sqrt{2mE}}{\hbar}a = n\pi \qquad (24.6)$$

Aside: The reader is reminded of the solution of the Newtonian wave equation.

And our solution is:

$$u = B\sin\left(\frac{n\pi}{a}x\right) \qquad (24.7)$$

With energies (energy eigenvalues):

$$E_n = \frac{\hbar^2}{2m}\frac{n^2\pi^2}{a^2} \qquad (24.8)$$

Aside: We remind the reader that Bohr calculated the energy levels of the hydrogen atom to be:

$$E_n^{\text{Hydrogen atom}} = -\frac{\hbar^2}{2m}\frac{1}{a_0^2}\frac{1}{n^2} \qquad (24.9)$$

The minus sign is from the arbitrary way we labeled the potential energy.

We see that the energy levels of the infinite square well increase as n^2. This contrasts with the energy levels of real atoms which increase as $\frac{1}{n^2}$.

We see that there are discreet energy levels in the infinite square well. The graphs of those solutions, when $a = 6$, are:

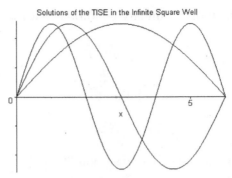

Solutions of the TISE in the Infinite Square Well

We see that it is the imposition of the boundary conditions that leads to the discreetness of the allowed energies. We see that each allowed energy corresponds to a different frequency of the standing waves between the walls of the infinite square well.

This allows us to picture an electron "jumping" from one energy to another. We can interpret this as the electron wave being a circular standing wave "sitting" around the nucleus of the hydrogen atom. The circumference of the standing wave is an integer number of electron wavelengths, and the number of wavelengths corresponds to the energy of the orbit.

The reader will have noticed that, in the case of the infinite square well potential, we have chosen particular solutions and discarded other solutions. The solutions we have discarded either lead to qualitatively the same consequences – energy quantitisation – or they are not suitable as wavefunctions because they are not square integrable. There are non-separable solutions to the TDSE that we also disregard. We justify this disregarding of solutions by saying that the TDSE is a model of reality and that, when we take the solutions we want, it works magnificently. We are really just picking out the basis solutions.

Because the walls of the infinite square well are infinite, there is no penetration of the walls by the particle. These walls are infinitely hard. If the walls of the well were not infinitely hard, they would give a little when hit by a rubber ball, we see something like this in the next example of a potential.

> **Technical note:** The bound eigenfunctions of a potential well are not a complete set of functions. However, the bound and unbound eigenfunctions of a potential well, when taken together, are a complete set of functions.

CHAPTER 25

THE STEP POTENTIAL

25.1 REFLECTION AND TRANSMISSION

If you stand inside a room alongside a loud radio, you will find that some part of the sound waves are reflected from the walls of the room and you will find that some part of the sound waves penetrate through the walls of the room and are heard by your neighbour. Waves can both go through a wall and bounce back from a wall. If you switch on a light within the room, you will see the light bulb reflected from the glass of the window. At the other side of the glass, your neighbour will see the light bulb through the glass. Light waves can both go through a pane of glass and bounce back from that pane of glass. Given that particles are waves, it ought then to be no surprise that particles can both bounce from walls and go through walls. This "going through walls" property of sound waves has been known since the Neolithic[1], so it is quite anachronistic that it be referred to as quantum tunneling; none-the-less it is so called.

We are shortly to look at a step potential. This is a model of a particle scattering from an impact with an atomic nucleus of other

---

[1.] See "Quantum Tunneling and Noisy Neighbours" by Gus Grit the Ancient Brit.

such step-like potential. We will find that part of the wavefunction penetrates the potential and part is reflected from the potential.

25.2 THE POTENTIAL STEP

The potential step is an idealized 1-dimensional potential energy. To the left of the origin, the potential is zero; to the right of the origin, the potential is a finite amount. We have:

$$V(x) = 0 \quad : \quad x < 0$$
$$V(x) = V \quad : \quad x > 0$$

(25.1)

Of course, a realistic step potential would not have the abrupt discontinuity at $x = 0$, but, because we cannot simply solve the realistic step potential, we solve the idealized, discontinuous step potential.

A Realistic Step Potential

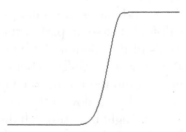

We take it that a particle which "feels" the force is moving across the page from left to right or from right to left. Associated with this potential, as with all potentials in both classical physics and quantum mechanics, there is a force given by:

$$F = -\frac{\partial V}{\partial x}$$

(25.2)

Both classically and in quantum mechanics, this force acts upon a particle to push it leftward. In a realistic potential, which would not be an infinitely steep step, a particle would "feel" this force in the vicinity of the origin. Away from the origin, both to the left and to the right, the particle would effectively behave as a free particle. The energy of the particle is the sum of its kinetic energy and its potential energy; both of these are functions of the particle's position, x. We have, both classically and quantum mechanically:

$$E_{Particle} = K.E.(x) + P.E.(x) = T(x) + V(x) \qquad (25.3)$$

25.3 CLASSICALLY

A. We first consider the situation classically when the total energy of the particle is greater than the height of the potential energy barrier, $E_{Particle} > V$. The particle approaches the potential step from the left. Before it "feels" the potential, the particle's potential energy is zero but its kinetic energy is T_0. As the particle passes through the potential step, the particle's velocity lessens and its kinetic energy is gradually converted into potential energy until, when it has passed through the barrier it has an amount of potential energy equal to V and an amount of kinetic energy equal to $T_0 - V > 0$. The thing to note is that the whole of the particle passes through the potential step. We call this "Total Transmission."

B. We next consider the situation classically when the total energy of the particle is less than the height of the potential energy barrier, $E_{Particle} < V$. The particle approaches the potential step from the left. Before it "feels" the potential, the particle's potential energy is zero but its kinetic energy is T_0. As the particle passes through the potential step, the particle's velocity lessens and its kinetic energy is gradually converted into potential energy until there is no more kinetic energy left and the particle "bounces" off the potential step and its motion is reversed. During this reversed motion, the potential energy of the particle is converted to kinetic energy until there is no more potential energy left and the particle has the same amount of kinetic energy as it started with but its motion is reversed. The thing to note is that the whole of the particle bounces off the potential step. We call this "Total Reflection."

25.4 QUANTUM MECHANICALLY

The quantum mechanical treatment leads to phenomena that are different from the total transmission and total reflection of the classical understanding.

The step potential is really two different potentials that meet at the origin. These potentials are independent of time, and so, in a quantum mechanical consideration of the situation, the motion of the particle is determined by the energy eigenvalue equation that is the time independent Schrödinger equation, TISE. Since there are two 1-dimensional potentials, there are two 1-dimensional time independent Schrödinger equations. We have:

$$-\frac{\hbar^2}{2m}\frac{\partial^2 u_E}{\partial x^2} + V u_E = E_n u_E \tag{25.4}$$

A little manipulation:

$$\frac{\partial^2 u_E}{\partial x^2} + \frac{2m}{\hbar^2}(E_n - V)u_E = 0 \tag{25.5}$$

There are two case when $V = V$ and when $V = 0$. Either way, we have to find u_E such that when it is differentiated twice it is equal to a multiple of itself. By looking at this equation for long enough, we see that the two cases have the solutions:

$$u_{x<0} = Ae^{i\sqrt{\frac{2mE_n}{\hbar^2}}x} + Be^{i\sqrt{\frac{2mE_n}{\hbar^2}}(-x)}$$

$$u_{x>0} = Ce^{i\sqrt{\frac{2m(E_n-V)}{\hbar^2}}x} + De^{i\sqrt{\frac{2m(E_n-V)}{\hbar^2}}(-x)} \tag{25.6}$$

These are the free particle solutions (sums of basis functions). Of course they are, the uniform potentials are each effectively a zero potential. In fact, there are an infinite number of solutions in each case that can be formed as linear combinations of the appropriate solution given above.

The two wavefunctions have two parts. One part corresponds to x in the positive direction, and the other part corresponds to x in the negative direction. These correspond to the particle traveling from left to right and to the particle traveling from right to left.

For a solution to be a wavefunction, it must have particular properties such as being square integrable, continuity, and differentiability. These properties mean that the wavefunction must be such that:

i.   The wavefunction is continuous at the boundary between the two potentials.

$$\psi_{Left}(0) = \psi_{Right}(0) \tag{25.7}$$

ii.   The first derivative of the wavefunction is continuous at the boundary between the two potentials.

$$\frac{\partial}{\partial x}\psi_{Left}(0) = \frac{\partial}{\partial x}\psi_{Right}(0) \tag{25.8}$$

> **Technically:** Since the change in potential at the step is finite and $u_E(0)$ is finite (a general boundary condition of the time independent Schrödinger equation), the time independent Schrödinger equation says that $\frac{\partial^2 u_E}{\partial x^2}$ is finite. This means that both u_E and $\frac{\partial u_E}{\partial x}$ are continuous across the potential.

25.5   PARTICLE INCIDENT FROM THE LEFT

This scenario is described by solutions of the TISE applied to the left part of the potential step, $V = 0$. Such solutions have both incident and reflected cases. Such a solution is a sum of the two possibilities of the form:

$$
\begin{aligned}
u_{Left} &= \text{Incident} + \text{Reflected} \\
&= Ae^{i\sqrt{\frac{2mE_n}{\hbar^2}}x} + Be^{i\sqrt{\frac{2mE_n}{\hbar^2}}(-x)}
\end{aligned}
\tag{25.9}
$$

Where $\{A, B\}$ are constants whose moduli measure the intensity of the incident wave and the intensity of the reflected wave respectfully. There could also be a transmitted wave traveling from left to right, $V = V$. The direction of travel means we can have only:

$$\text{Transmited } + \text{ Reflected}$$

$$u = Ce^{i\frac{\sqrt{2m(E_n-V)}}{\hbar}x} + De^{i\frac{\sqrt{2m(E_n-V)}}{\hbar}(-x)} \tag{25.10}$$

Where C is a constant whose modulus measures the intensity of the transmitted wave. For $E_n < V$, the reflected part of the wave in the $x > 0$ region is not square integrable, and so it is not a wavefunction and so we set $D = 0$. Now, we need the wavefunction to be continuous across the boundary between the two potentials, that is at $x = 0$. We need:

$$Ae^{i\sqrt{\frac{2mE_n}{\hbar^2}}x} + Be^{-i\sqrt{\frac{2mE_n}{\hbar^2}}x} = Ce^{i\frac{\sqrt{2m(E_n-V)}}{\hbar}x} \tag{25.11}$$

At $x = 0$:

$$A + B = C \tag{25.12}$$

We also need the differential of the wavefunction to be continuous at $x = 0$. This leads to:

$$i\sqrt{\frac{2mE_n}{\hbar^2}}Ae^{i\sqrt{\frac{2mE_n}{\hbar^2}}x} - i\sqrt{\frac{2mE_n}{\hbar^2}}Be^{-i\sqrt{\frac{2mE_n}{\hbar^2}}x} = i\frac{\sqrt{2m(E_n-V)}}{\hbar}Ce^{i\frac{\sqrt{2m(E_n-V)}}{\hbar}x} \tag{25.13}$$

$$\sqrt{E_n}A - \sqrt{E_n}B = \sqrt{(E_n-V)}C$$

We have three constants but only two equations, but we can arbitrarily normalize one of the constants to unity. We put $A = 1$, and this gives:

$$1 + B = C$$
$$\sqrt{E_n}(1-B) = \sqrt{(E_n-V)}C \tag{25.14}$$

Leading to:

$$B = \frac{\sqrt{E_n} - \sqrt{(E_n-V)}}{\sqrt{E_n} + \sqrt{(E_n-V)}} \tag{25.15}$$

$$C = \frac{2\sqrt{E_n}}{\sqrt{E_n} + \sqrt{(E_n-V)}} \tag{25.16}$$

We now have three numbers measuring the intensities of the incident wave, $A = 1$, the reflected wave, B, and the transmitted wave, C. If $B = 0$, there is no reflected wave. If $C = 0$, there is no transmitted wave.

The transmitted wave is non-zero unless $E_n = 0$. The reader might think this has happened because we fed the transmitted wave into the equations. We assumed that there was no wave traveling from right to left in the $x > 0$, and then we simply fed in all possibilities. We then applied the continuity relations across the boundary. It could have happened that the continuity conditions meant $C = 0$ and there was no transmitted wave, but it did not.

Looking at the constants $\{B, C\}$ in (25.16), we see that if $E_n \gg V$ then $V \sim 0$, the intensity of the reflected wave, $|B|^2$, would be zero and there would be no reflected wave and the whole wave would be transmitted, as we would expect. If $E_n \ll V$, the intensity of the transmitted wave, $|C|^2$, tends to zero, and there would be no transmitted wave, as we would expect.

This is very different from the total transmission or total reflection of a classical particle (bouncing ball). In quantum mechanics, because we are treating particles as waves, we have partial transmission and partial reflection.

25.6 REFLECTED NEUTRONS

There is an interesting quantum mechanical phenomenon associated with low energy (slow moving) particles, like neutrons, that are incident upon a highly attractive potential such as the surface of an atomic nucleus. Classically, the slow moving particle would accelerate toward the nucleus, but, in fact, slow moving neutrons are reflected by the attractive potential rather than attracted by it. A highly attractive potential is $-V$. Putting this into the equations (25.16) gives:

$$B = \frac{\sqrt{E_n} - \sqrt{(E_n + V)}}{\sqrt{E_n} + \sqrt{(E_n + V)}}$$

$$C = \frac{2\sqrt{E_n}}{\sqrt{E_n} + \sqrt{(E_n + V)}}$$

(25.17)

For $|E_n| \ll |V|$, we have $C \sim 0$, and so there is negligible transmission – the neutron is reflected from the attractive nucleus and not transmitted into the nucleus. This is experimental confirmation of the theory outlined above and experimental disproof of the classical theory of particles.

Aside: Neutrons were discovered by James Chadwick in 1932[2].

25.7 QUANTUM TUNNELING

So, the particle, which is a standing wave, is split into three pieces by the step in potential. One piece is approaching the step, another piece is reflected from the step, and another piece is transmitted through the step. The wavefunction of the particle is the sum of the three eigenfunctions that describe the incident, reflected, and transmitted waves. So, where is the particle?

The probability of finding the particle at a given point, $x > 0$, is given by:

$$P_{x>0} = |u_E|^2 = \left| Ce^{i\sqrt{\frac{2m(E_n-V)}{\hbar^2}}x} \right|^2 \tag{25.18}$$

For $E_n < V$, $\sqrt{\frac{2m(E_n-V)}{\hbar^2}}$ is imaginary, and we have:

$$P_{x>0} = |C|^2 e^{-2\sqrt{\frac{2m(V-E_n)}{\hbar^2}}x} \tag{25.19}$$

This probability decays exponentially as x increases.

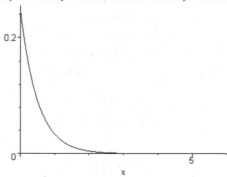

Decay of Probability of Transmission into Classically Forbidden Reç

So there is negligible probability (intensity) of the particle at $x > 2$, but some intensity (probability) of the particle at $0 < x < 1$.

---

[2] J. Chadwick, Nature Feb 27 1932.

This means, that for any particle/wave with any energy, E_n, some of the particle will be inside the potential step – the intensity field (probability field) is not zero inside the potential step. For $E_n < V$, this would be impossible for a classical particle. A classical particle would just bounce off the potential step. It happens because the particle is a wave – it happens with sound waves, does it not? We refer to this penetration of a particle into regions that a classical particle could not reach as quantum tunneling. If we were to measure the position of the particle, the wavefunction would collapse into one of the three eigenfunctions corresponding to incident, reflected, and transmitted waves. Sometimes, it will collapse into the transmitted eigenfunction, and we will measure the position of the particle as being inside the classically forbidden region. The particle has "tunnelled" into a region where is classically cannot be.

25.8   A NARROW POTENTIAL BARRIER

With the above we have all the ingredients we need to analyse a narrow potential barrier. There will be both incident and reflected waves in all regions of the barrier. There will be exponential decay through the barrier in the direction of the incident wave and exponential decay in the direction of the reflected wave. We show a wave incident from the left; we do not show any reflected waves.

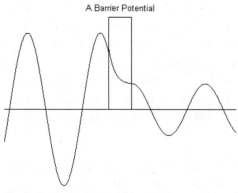

A Barrier Potential

We see that a particle, because it is a wave, with less kinetic energy than the potential barrier, can possibly penetrate the barrier. This can never happen classically. This is quantum tunneling. If we calculate the intensity of the wave (the probability that the particle will be observed) at each point along the x-axis, we get:

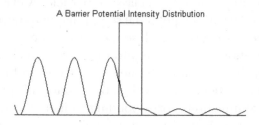

A Barrier Potential Intensity Distribution

25.9 ALPHA PARTICLE EMISSION

Many atomic nuclei radioactively decay by emitting an alpha particle (helium nucleus). Each particular nucleus has a characteristic half-life. There is an inverse correlation between the energy of the emitted alpha particle and the half-life of the particular nucleus as was noted by Rutherford in 1907. This dependence is remarkable in that the energies of emitted alpha particles range from 4 Mev to 9 Mev but the half-lives range from less than a microsecond to more than 10^{11} years.

In the view of classical physics, alpha particle emission is impossible. Experiments by Rutherford and Royds[3] in 1907 using alpha particles with energies of 8.7 Mev had demonstrated a Coulomb potential barrier around uranium nuclei of at least 8.7 Mev, and so it would be impossible for an alpha particle with only 4.2 Mev to pass through this potential barrier. None-the-less, uranium nuclei emit alpha particles with energies of 4.2 Mev. This is an example of quantum mechanical tunneling. It was first realized to be such in 1928[4] by Gamow (1904–1968) and independently by Gurney and Condon also in 1928.[5]

[3.] E Rutherford & T Royds (1908) Spectrum of the radium emanation : Philosophical Magazine, Series 6, Vol. 16, pgs 313–317.

[4.] Z. Physik 5.1 204 (1928) G Gamow Zur Quantentheorie des Atomkernes.

[5.] Gurney R.W. & Condon E.U. (1928) Quantum Mechanics and Radioactive Distintergration : Nature 122.439.

26

THE 3-DIMENSIONAL BOX AND DEGENERACY

Schrödinger's time independent equation in 3-dimensions is:

$$-\frac{\hbar^2}{2m}\nabla^2 u(r) + V(r)u(r) = Eu(r) \qquad (26.1)$$

The form of the potential, $V(r)$ will dictate which is the most suitable co-ordinate system in which to write the TISE. There are eleven different co-ordinate systems in which the 3-dimensional TISE, (26.1), is separable. The equation is separable only if the potential can be written as a sum of three terms; each of which is a function of only one co-ordinate. For simplicity, we assume the potential is of the form:

$$V(r) = V_x(x) + V_y(y) + V_z(z) \qquad (26.2)$$

In this case, it is possible to find solutions of the form:

$$u(r) = X(x)Y(y)Z(z) \qquad (26.3)$$

A little algebraic manipulation and the realization that the co-ordinates are mutually independent leads to the three equations:

$$-\frac{\hbar^2}{2m}\frac{\partial^2}{\partial x^2}X(x) + V_x X(x) = E_x X(x)$$

$$-\frac{\hbar^2}{2m}\frac{\partial^2}{\partial y^2}Y(y) + V_y Y(y) = E_y Y(y) \qquad (26.4)$$

$$-\frac{\hbar^2}{2m}\frac{\partial^2}{\partial z^2}Z(z) + V_z Z(z) = E_z Z(z)$$

The total energy is simply the sum of the three separate energies:

$$E = E_x + E_y + E_z$$

(26.5)

The eigenfunctions are the products of the individual co-ordinate eigenfunctions:

$$u_{n_x n_y n_z}(x,y,z) = X_{n_x}(x)Y_{n_y}(y)Z_{n_z}(z)$$

(26.6)

We apply the 3-dimensional Schrödinger equation to a particle in a 3-dimensional box of size {2a,2b, 2c} with potential given by:

$$
\begin{array}{llll}
V_x(x) = 0 & : & -a < x < a & \& \quad V_x(x) = \infty \quad : \quad |x| \ge a \\
V_y(y) = 0 & : & -b < y < b & \& \quad V_y(y) = \infty \quad : \quad |y| \ge b \\
V_z(z) = 0 & : & -c < x < c & \& \quad V_z(z) = \infty \quad : \quad |z| \ge c
\end{array}
$$

(26.7)

We have already solved these equations when we dealt with the 1-dimensional infinite square well. The solutions are:

$$X(x) = a^{-\frac{1}{2}} \cos\left(\frac{n_x \pi}{2a} x\right) \quad : \quad n_x \text{ is even}$$

$$X(x) = a^{-\frac{1}{2}} \sin\left(\frac{n_x \pi}{2a} x\right) \quad : \quad n_x \text{ is odd}$$

(26.8)

With similar solutions for the other co-ordinates. The energies are:

$$E_x = \frac{\pi^2 \hbar^2}{8ma^2} n_x^2, \quad E_y = \frac{\pi^2 \hbar^2}{8mb^2} n_y^2, \quad E_z = \frac{\pi^2 \hbar^2}{8mc^2} n_z^2$$

(26.9)

And the total energy is:

$$E = \frac{\hbar^2 \pi^2}{8m}\left(\frac{n_x^2}{a^2} + \frac{n_y^2}{b^2} + \frac{n_z^2}{c^2}\right)$$

(26.10)

26.1 DEGENERACY

There is a reason that text books on quantum mechanics include a 3-dimensional box solution; it is because the 3-dimensional box allows the authors to demonstrate degeneracy. Assume that the width and depth of the box are the same – it is a square box. The total energy is then:

$$E = \frac{\hbar^2 \pi^2}{8m}\left(\frac{n_x^2}{a^2} + \frac{n_y^2}{a^2} + \frac{n_z^2}{c^2}\right)$$ (26.11)

In these circumstances, we get the same total energy for different values of $\{n_z, n_y, n_z\}$. For example, the total energy is the same for the two cases $\{n_x = 1, n_y = 2, n_z = 1\}$ and $\{n_x = 2, n_y = 1, n_z = 1\}$. These two cases correspond to two different eigenfunctions, and so we have two different eigenfunctions with the same energy. This is two different eigenfunctions with the same eigenvalue. This is called degeneracy.

27

PARITY

Undoubtedly, the reader is now sitting very comfortably, and so we shall continue. Within the complex number division algebra $\mathbb{C}$, there are two basic trigonometric functions. These are the cosine function and sine function that govern rotation in the complex plane.

$$\mathbb{C} \equiv \begin{bmatrix} r & 0 \\ 0 & r \end{bmatrix} \begin{bmatrix} \cos\theta & \sin\theta \\ -\sin\theta & \cos\theta \end{bmatrix} \qquad (27.1)$$

It is the necessary nature of trigonometric functions that they exhibit the symmetry and the anti-symmetry of the geometric space of which they are a part. We have:

$$\begin{aligned} Symmetry: & \quad \cos(-\theta) = \cos(\theta) \\ Anti-symmetry: & \quad \sin(-\theta) = -\sin(\theta) \end{aligned} \qquad (27.2)$$

Waves are connected to the complex numbers, $\mathbb{C}$, through the wavy nature of the trigonometric functions of that space.

Aside: It is a consequence of the symmetry and anti-symmetry of the trigonometric functions that a clockwise rotation through θ followed by an counterclockwise rotation through θ takes us back to where we started. Suppose that the sine function was symmetric like the cosine function. Let us rotate first in the clockwise direction then in the counterclockwise direction. Watch the minus signs:

$$
\begin{bmatrix} \cos\theta & \sin\theta \\ -\sin\theta & \cos\theta \end{bmatrix}
\begin{bmatrix} \cos(-\theta) & \sin(-\theta) \\ -\sin(-\theta) & \cos(-\theta) \end{bmatrix}
$$

$$
= \begin{bmatrix} \cos\theta & \sin\theta \\ -\sin\theta & \cos\theta \end{bmatrix}
\begin{bmatrix} \cos(\theta) & \sin(\theta) \\ -\sin(\theta) & \cos(\theta) \end{bmatrix}
\qquad (27.3)
$$

$$
= \begin{bmatrix} \cos(2\theta) & \sin(2\theta) \\ -\sin(2\theta) & \cos(2\theta) \end{bmatrix}
$$

We do not return to where we started. Only because the cosine function is symmetric and the sine function is anti-symmetric does a clockwise rotation through θ followed by an counterclockwise rotation through θ take us back to where we started.

$$
\begin{bmatrix} \cos\theta & \sin\theta \\ -\sin\theta & \cos\theta \end{bmatrix}
\begin{bmatrix} \cos(-\theta) & \sin(-\theta) \\ -\sin(-\theta) & \cos(-\theta) \end{bmatrix}
$$

$$
= \begin{bmatrix} \cos\theta & \sin\theta \\ -\sin\theta & \cos\theta \end{bmatrix}
\begin{bmatrix} \cos(\theta) & -\sin(\theta) \\ \sin(\theta) & \cos(\theta) \end{bmatrix}
\qquad (27.4)
$$

$$
= \begin{bmatrix} 1 & 0 \\ 0 & 1 \end{bmatrix}
$$

The graph of the cosine function is "balanced" across the origin:

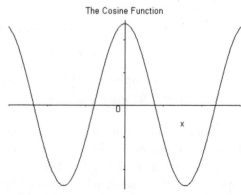

The Cosine Function

Of course it is because the cosine is the projection from the unit circle on to the real axis. The graph of the sine function is exactly "unbalanced" across the origin:

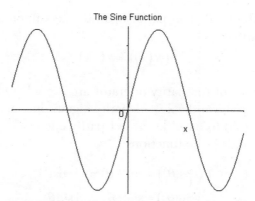

The Sine Function

We refer to the cosine function as being an even function and the sine function as being an odd function. We have:

$$\text{Even function}: \quad f(x) = f(-x)$$
$$\text{Odd function}: \quad f(x) = -f(x) \tag{27.5}$$

Now, since a wavefunction is a sum of waves, it will be a sum of symmetric waves based on the cosine function and a sum of anti-symmetric waves based on the sine function. For any wavefunction, u_n, we have the even part of the wavefunction being given by:

$$u_{even} = \frac{1}{2}\left[u_n(x) + u_n(-x)\right] \tag{27.6}$$

And we have the odd part of the wavefunction being given by:

$$u_{odd} = \frac{1}{2}\left[u_n(x) - u_n(-x)\right] \tag{27.7}$$

27.1 THE PARITY OPERATOR

We refer to this oddness and evenness, this anti-symmetry and symmetry, this "balance" and "unbalance" as parity. We say that a particular wavefunction is of even parity or that it is of odd parity, or we split it into its even part and its odd part.

We take the view that there is a parity operator, $\hat{P}$, which we have to be careful to distinguish from the momentum operator, $\hat{P}$, or the probability, P. The parity operator acts upon a function to reverse its parity:

$$\hat{P}(\psi(x)) = \psi(-x) \qquad (27.8)$$

The eigenvalues of the parity operator are $\lambda = \pm 1$. Odd wavefunctions are said to have the $\lambda = -1$ parity eigenvalue, and even wavefunctions are said to have the $\lambda = +1$ parity eigenvalue. We see this with the sine and cosine functions:

$$\hat{P}(\cos(\theta)) = \cos(-\theta) = +1\cos\theta$$
$$\hat{P}(\sin(\theta)) = \sin(-\theta) = -1\sin\theta \qquad (27.9)$$

For symmetric potentials, the parity operator commutes with the Hamiltonian (energy operator).

27.2 THE SYMMETRIC POTENTIAL WELL

We consider a potential well that is symmetric about the origin.

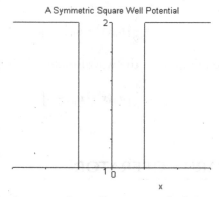

A Symmetric Square Well Potential

Within the well, there are bound states with discreet energy levels. Above the well, there are continuous free states. However, because the walls of this potential well are not infinite, the bound states will "leak" out of the well into the potential where they could not be classically. Because the potential is symmetric about the origin, there will be bound states that are symmetric (based on the cosine function). More surprisingly, there are also bound anti-symmetric states (based on the sine function). The graphs of the eigenfunctions of this well are:

An Even Eigenfunction in the Symmetric Potential

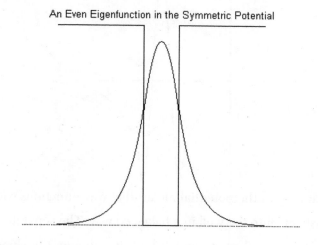

An Odd Eigenfunction in the Symmetric Potential

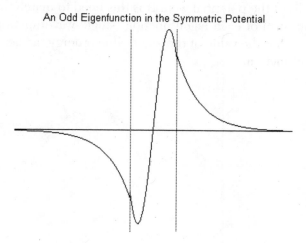

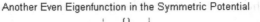

Another Even Eigenfunction in the Symmetric Potential

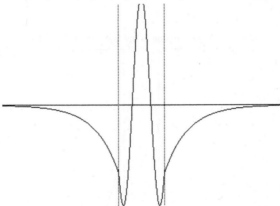

The symmetry of the potential means that wavefunctions within the well must "fit" into the well in a balanced way. The separate parts of the wavefunctions have to match both in value and in slope $\left(\dfrac{\partial y}{\partial x}\right)$ at the walls of the potential well. It is this need to match that determines the form of each eigenfunction (single wavefunction) in the well and with it the value of the eigenvalue (energy) associated with that eigenfunction.

CHAPTER 28

THE SIMPLE HARMONIC OSCILLATOR

Until now, we have dealt with potentials that are unreal such as the infinite square well or the infinitely steep potential step. Such potentials correspond to infinitely strong forces and discontinuities that do not occur in nature. We have used these potentials to demonstrate different aspects of quantum mechanics because the Schrödinger equation with such potentials is easy to solve. To do physics in the real world, we need to be able to solve the Schrödinger equation with real potentials. In almost all cases, this is very difficult and we have to resort to numerical approximation. However, in the case of the simple harmonic oscillator, SHO, potential we can solve the Schrödinger equation without resorting to numerical methods. This is very useful because all real potential wells approximate the potential well of the simple harmonic oscillator close to the bottom of the well. In stable physical systems, a particle will oscillate about the bottom of a potential well in a way that is similar to how a particle behaves in a simple harmonic oscillator potential. Consider an atom attached to other atoms to form a molecule; there is an attractive electrostatic force that holds the atom close to the other atoms, and there is a repulsive nuclear force that prevents the atom from getting too close to the other atoms. The atom is in a potential well, and it vibrates around the bottom of this potential well in a way that is approximated by a particle in the simple harmonic oscillator potential.

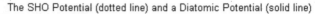

The SHO Potential (dotted line) and a Diatomic Potential (solid line)

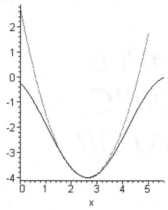

The mathematics of this similarity is in the Taylor expansion of the potential, $V(x)$. That Taylor expansion is:

$$V(x) = V(x_0) + (x - x_0)\frac{\partial V}{\partial x} + \frac{1}{2}(x - x_0)^2 \frac{\partial^2 V}{\partial x^2} + \dots \qquad (28.1)$$

The leftmost term is a constant that can be ignored within a potential. The middle term is zero at the bottom of the well because the slope, $\frac{\partial V}{\partial x}$, is zero there. The first and most significant term at the bottom of the well is of the form:

$$\frac{1}{2}(x - x_0)^2 \frac{\partial^2 V}{\partial x^2} \sim \frac{1}{2}Kx^2 \qquad : \qquad K = \frac{\partial^2 V}{\partial x^2} \qquad (28.2)$$

This is the form of the simple harmonic oscillator potential, $V = \frac{1}{2}Kx^2 = \frac{1}{2}m\omega^2 x^2$. We have used this term to draw the dotted graph above.

28.1 THE TISE OF THE SIMPLE HARMONIC OSCILLATOR

Because the potential is independent of time, we use the time independent Schrödinger equation, TISE. This assumes, usually without stating it, that the solution of the time dependent Schrödinger equation, TDSE, is separable, and any solutions we find will be separable solutions. When we have found separable solutions, ψ, of the time independent Schrödinger equation TISE, we can turn them into solutions of the time dependent Schrödinger equation, TDSE, by multiplying them by $e^{-i\frac{Et}{\hbar}}$.

The energy eigenvalue equation which is the TISE of the simple harmonic operator, SHO, is:

$$\left(-\frac{\hbar^2}{2m}\frac{\partial^2}{\partial x^2}+V(x)\right)u(x)=Eu(x) \tag{28.3}$$

Re-arranging:

$$-\frac{\hbar^2}{2m}\frac{\partial^2 u(x)}{\partial x^2}+\frac{1}{2}m\omega^2 x^2 u(x)=Eu(x)$$

$$\frac{\hbar}{m\omega}\frac{\partial^2 u(x)}{\partial x^2}+\left(\frac{2E}{\omega\hbar}-\frac{m\omega}{\hbar}x^2\right)u(x)=0 \tag{28.4}$$

With $y=\sqrt{\frac{m\omega}{\hbar}}x$ and $\lambda=\frac{2E}{\hbar\omega}$, this becomes:

$$\frac{\partial^2 u(y)}{\partial y^2}+(\lambda-y^2)u(y)=0 \tag{28.5}$$

In this book, we will solve this equation, (28.5), in two different ways. We do not need to solve the SHO TISE in two ways, and most text books solve it in only one way, but we opine that the reader will very much enjoy seeing both methods. It is also true that we are able to show important results of quantum mechanics by solving the SHO TISE by two different methods.

28.2 SOLUTION 1 (FROBENIUS METHOD)

We need solutions that are square integrable. This means that we need solutions whose graphs have a non-infinite area between them and the axis. This means that we need solutions that do not diverge for very large, negative or positive, values of the variable. We start by considering the behavior of the equation above, (28.5), when y is large. When y is large, we can neglect λ, and the form of the solution of the above equation, (28.5), will approach the form of the solution of:

$$\frac{\partial^2 u(y)}{\partial y^2} - y^2 u(y) = 0 \qquad (28.6)$$

We try a solution of the form:

$$u = e^{\pm \frac{1}{2} y^2} \qquad (28.7)$$

Immediately, we see that we must reject the positive solution because it diverges for large y and therefore cannot be square integrable (normalizable). The negative solution is:

$$u = e^{-\frac{1}{2} y^2} \qquad (28.8)$$

This implies:

$$\frac{\partial^2 u}{\partial y^2} = \left(-1 + y^2\right) e^{-\frac{1}{2} y^2} \qquad (28.9)$$

We see, by luck, we have a solution of the above equation, (28.5), if:

$$\lambda = 1 = \frac{2E}{\hbar \omega} \qquad (28.10)$$

This implies:

$$E = \frac{\hbar \omega}{2} \qquad (28.11)$$

We have our first eigenvalue. We also have that the behavior as $y \rightarrow \infty$ is given by $e^{-\frac{1}{2} y^2}$. We therefore look for solutions of the form:

$$u = H(y) e^{-\frac{1}{2} y^2} \qquad (28.12)$$

The $H(y)$ stands for Hermite polynomials – it is not the Hamiltonian. The $e^{-\frac{1}{2}y^2}$ term decays very rapidly. Only if the $H(y)$ term increases more rapidly than the $e^{-\frac{1}{2}y^2}$ term decreases will the solution not be square integrable. Remember, the linear space of square integrable functions, $L^2(\mathbb{R}^3)$, is also the set of possible wavefunctions.

With solution $u = H(y)e^{-\frac{1}{2}y^2}$, the SHO TISE is:

$$\frac{\partial^2 H(y)e^{-\frac{1}{2}y^2}}{\partial y^2} + \left(\lambda - y^2\right)H(y)e^{-\frac{1}{2}y^2} = 0$$

$$e^{-\frac{1}{2}y^2}\left(\frac{\partial^2 H(y)}{\partial y^2} - 2y\frac{\partial H(y)}{\partial y} + y^2 H(y) - H(y)\right) \qquad (28.13)$$

$$+\left(\lambda - y^2\right)H(y)e^{-\frac{1}{2}y^2} = 0$$

$$\left(\frac{\partial^2 H(y)}{\partial y^2} - 2y\frac{\partial H(y)}{\partial y} + (\lambda - 1)H(y)\right) = 0$$

This is known as Hermite's equation. Insisting upon square integrability leads to an infinite number of solutions of Hermite's equation that are polynomials in y. These polynomials are known as Hermite polynomials. The first few Hermite polynomials are:

$$H_0 = 1$$
$$H_1 = 2y$$
$$H_2 = 4y^2 - 2$$
$$H_3 = 8y^3 - 12y \qquad (28.14)$$
$$H_4 = 16y^4 - 48y^2 + 12$$
$$H_5 = 32y^5 - 160y^3 + 120y$$

...

The Hermite polynomials are generated by the recurrence relation:

$$H_n = 2yH_{n-1} - 2(n-1)H_{n-2} \qquad (28.15)$$

There is one eigenfunction of the SHO for each of these Hermite polynomials.

Throwing in a normalizing real constant, N_n, the wavefunctions (eigenfunctions) that are square integrable solutions of the TISE are:

$$u_n = N_n e^{-\frac{1}{2}y^2} H_n(y)$$
$$= N_n e^{-\frac{m\omega}{2\hbar}x^2} H_n\left(\sqrt{\frac{m\omega}{\hbar}}x\right)$$

(28.16)

Since the SHO potential well is symmetrical, the wavefunctions will have either even or odd parity. The graphs of the first few eigenfunctions are:

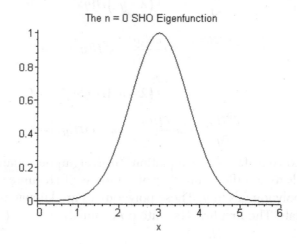

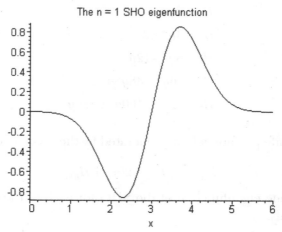

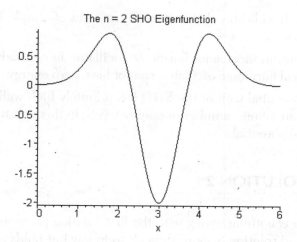

The n = 2 SHO Eigenfunction

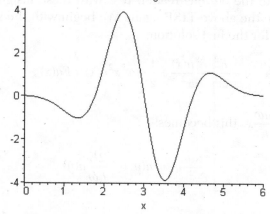

The n = 3 SHO eigenfunction

Calculations put the normalizing constant to be:

$$N_n = \sqrt{\frac{1}{2^n n!}} \sqrt[4]{\frac{m\omega}{\pi\hbar}} \qquad (28.17)$$

And the energy eigenvalues are given by:

$$E_n = \left(n + \frac{1}{2}\right)\hbar\omega \qquad (28.18)$$

When $n = 0$, we have the lowest energy level, $E_0 = \frac{1}{2}\hbar\omega$. Notice how the energy levels of the SHO, that is the energy eigenvalues of the SHO TISE, increase in lumps of $\hbar\omega$. This is just what Planck proposed. It indicates that matter is just a set of simple harmonic oscillators.

A quantum mechanics harmonic oscillator, in contradistinction to a classical harmonic oscillator, cannot have zero energy.

The potential well of the SHO has infinitely high walls, and so there are an infinite number of energy levels; in this way, it does not match real potentials.

28.3 SOLUTION 2

There is nothing wrong with the first solution presented above. This second solution is more difficult to follow but holds insights of great value to the curious reader. It is with these insights in mind that we solve the above TISE again. We begin with the same TISE that we had for the first solution:

$$-\frac{\hbar^2}{2m}\frac{\partial^2 u(x)}{\partial x^2} + \frac{1}{2}m\omega^2 x^2 u(x) = Eu(x) \tag{28.19}$$

With $y = \sqrt{\frac{m\omega}{\hbar}}x$, this becomes:

$$\frac{\partial^2 u(y)}{\partial y^2} - y^2 u(y) = -\frac{2E}{\hbar\omega}u(y) \tag{28.20}$$

28.4 THE FIRST BIT OF THE CALCULATION

Now consider the, not necessarily commutative, product of operators:

$$\left(\left(\frac{\widehat{\partial}}{\partial y}+\hat{y}\right)\left(\frac{\widehat{\partial}}{\partial y}-\hat{y}\right)\right)f(y) = \left(\frac{\widehat{\partial}}{\partial y}\frac{\widehat{\partial}}{\partial y}\right)f(y) - \left(\frac{\widehat{\partial}}{\partial y}\hat{y}\right)f(y)$$
$$+\left(\hat{y}\frac{\widehat{\partial}}{\partial y}\right)f(y) - \left(\hat{y}\hat{y}\right)f(y) \tag{28.21}$$
$$=\frac{\partial^2 f(y)}{\partial y^2} - \frac{\partial(y.f(y))}{\partial y} + y.\frac{\partial f(y)}{\partial y} - y^2 f(y)$$

$$= \frac{\partial^2 f(y)}{\partial y^2} - y \cdot \frac{\partial f(y)}{\partial y} - 1.f(y)$$

$$+ y \cdot \frac{\partial f(y)}{\partial y} - y^2 f(y) \qquad (28.22)$$

$$= \left(\frac{\widehat{\partial^2}}{\partial y^2} - \widehat{y^2} \right) f(y) - \widehat{1}.f(y)$$

Using this, we can rewrite the TISE, (28.20), as:

$$\left(\left(\frac{\widehat{\partial}}{\partial y} + \widehat{y} \right) \left(\frac{\widehat{\partial}}{\partial y} - \widehat{y} \right) \right) u(y) = \left(-\frac{2E}{\hbar\omega} - 1 \right) u(y) \qquad (28.23)$$

This equation is an eigenvalue equation. There are many eigenfunctions that satisfy this equation. We will distinguish the different eigenfunctions by attaching a subscripted number, n, $n + 1$, $n + 2$, ... to each of them.

The next step is to operate on both sides of the equation (28.23) with $\left(\frac{\widehat{\partial}}{\partial y} - \widehat{y} \right)$ to give:

$$\left(\left(\frac{\widehat{\partial}}{\partial y} - \widehat{y} \right) \left(\frac{\widehat{\partial}}{\partial y} + \widehat{y} \right) \left(\frac{\widehat{\partial}}{\partial y} - \widehat{y} \right) \right) u_n(y) = \left(-\frac{2E_n}{\hbar\omega} - 1 \right) \left(\frac{\widehat{\partial}}{\partial y} - \widehat{y} \right) u_n(y) \quad (28.24)$$

This equation is true if:

$$\left(\frac{\widehat{\partial}}{\partial y} - \widehat{y} \right) u_n(y) = 0 \qquad (28.25)$$

but the only solution of this is the non-square integrable $u_n = e^{+\frac{1}{2}y^2}$. This cannot be a wavefunction because it is not square integrable. (When we repeat this procedure in the second bit of this calculation, we will get a different answer that will be a wavefunction.)

The equation (28.25) is also true if this expression is equal to another eigenfunction (say $u_{n+1}(y)$). That is:

$$\left(\frac{\widehat{\partial}}{\partial y} - \widehat{y} \right) u_n(y) = u_{n+1}(y) \qquad (28.26)$$

Provided that:

$$\left(-\frac{2E_n}{\hbar\omega}-1\right)=\left(-\frac{2E_{n+1}}{\hbar\omega}+1\right)\Rightarrow E_{n+1}=E_n+\hbar\omega \qquad (28.27)$$

The reader should note that we have increased the energy eigenvalue by $\hbar\omega$ in going from E_n to E_{n+1}.

28.5 THE SECOND BIT OF THE CALCULATION

We now repeat the last few calculations with a few signs swapped around. By similar means as above, we calculate the "reversed" operator product:

$$\left(\left(\frac{\widehat{\partial}}{\partial y}+\hat{y}\right)\left(\frac{\widehat{\partial}}{\partial y}-\hat{y}\right)\right)f(y)=\left(\frac{\widehat{\partial^2}}{\partial y^2}-\widehat{y^2}\right)f(y)+\hat{1}.f(y) \qquad (28.28)$$

The reader is urged to look carefully at (28.22). There are signs different.

This leads to:

$$\left(\left(\frac{\widehat{\partial}}{\partial y}-\hat{y}\right)\left(\frac{\widehat{\partial}}{\partial y}+\hat{y}\right)\right)u(y)=\left(-\frac{2E}{\hbar\omega}+1\right)u(y) \qquad (28.29)$$

The next step is to operate on both sides of the equation (28.29) with $\left(\frac{\widehat{\partial}}{\partial y}+\hat{y}\right)$ to give:

$$\left(\left(\frac{\widehat{\partial}}{\partial y}+\hat{y}\right)\left(\frac{\widehat{\partial}}{\partial y}-\hat{y}\right)\left(\frac{\widehat{\partial}}{\partial y}+\hat{y}\right)\right)u_n(y)=\left(-\frac{2E_n}{\hbar\omega}+1\right)\left(\frac{\widehat{\partial}}{\partial y}+\hat{y}\right)u_n(y) \quad (28.30)$$

This equation is true if:

$$\left(\frac{\widehat{\partial}}{\partial y}+\hat{y}\right)u_n(y)=0 \qquad (28.31)$$

There is a sign difference from before.

This time, the only solution of this equation is the square integrable $u_n = e^{-\frac{1}{2}y^2}$. The reader should compare this with the earlier case. This is our ground state. We label it as:

$$u_0 = e^{-\frac{1}{2}y^2} \tag{28.32}$$

Associated with this eigenfunction is the ground state energy eigenvalue:

$$E_0 = \frac{1}{2}\hbar\omega \tag{28.33}$$

The equation (28.30) is also true if this expression is equal to another eigenfunction (say $u_{n-1}(y)$). That is:

$$\left(\frac{\widehat{\partial}}{\partial y} + \hat{y}\right)u_n(y) = u_{n-1}(y) \tag{28.34}$$

If

$$\left(-\frac{2E_n}{\hbar\omega} + 1\right) = \left(-\frac{2E_{n-1}}{\hbar\omega} - 1\right) \Rightarrow E_{n-1} = E_n - \hbar\omega \tag{28.35}$$

The reader should note that we have decreased the energy eigenvalue by $\hbar\omega$ in going from E_n to E_{n-1}.

After two phews worth of calculations, we combine the results to get the energy eigenvalues of the SHO, starting with the ground state, $n = 0$:

$$E_n = \left(n + \frac{1}{2}\right)\hbar\omega \quad : \quad n = 0,1,2,3,... \tag{28.36}$$

This is, of course, the same as the result of the first solution. So, why did we bother? Read on.

28.6 ANNIHILATION AND CREATION OPERATORS

Starting with the ground state energy eigenfunction, u_0, of the SHO, the operator, $\left(\frac{\partial}{\partial y} - y\right)$, will, subject to normalization, generate the successive SHO eigenfunctions.

$$u_1(y) = \left(\frac{\partial}{\partial y} - y\right) u_0(y)$$

$$= \left(\frac{\partial}{\partial y} - y\right) e^{-\frac{1}{2}y^2} \qquad (28.37)$$

$$= -2ye^{-\frac{1}{2}y^2} = -H_1 e^{-\frac{1}{2}y^2}$$

And:
$$u_2(y) = \left(\frac{\partial}{\partial y} - y\right) u_1(y)$$

$$= \left(\frac{\partial}{\partial y} - y\right)\left(-2ye^{-\frac{1}{2}y^2}\right) \qquad (28.38)$$

$$= \left(4y^2 - 2\right) e^{-\frac{1}{2}y^2}$$

$$= H_2 e^{-\frac{1}{2}y^2}$$

And so on. This again corresponds to the eigenfunctions we calculated in the first solution.

The operator $\left(\frac{\partial}{\partial y} + y\right)$ will, subject to normalization, spit out the eigenfunction immediately below the eigenfunction upon which it operates.

$$u_1(y) = \left(\frac{\partial}{\partial} + y\right) u_2(y)$$

$$= \left(\frac{\partial}{\partial} + y\right)\left(4y\ -2\right) e^{-} \qquad (28.39)$$

$$(8y)e^{-}$$

$$H\ e^{-}$$

And so on.

The two operators:

$$\hat{a}^\dagger = \left(\frac{\partial}{\partial y} - y\right)$$

$$\hat{a} = \left(\frac{\partial}{\partial y} + y\right) \qquad (28.40)$$

are creation and annihilation operators. The operator $\hat{a}^{\dagger}$ moves the SHO into the next higher energy level. In doing this, it must increase the energy by $\hbar\omega$; it must create energy. For this reason, the operator $\hat{a}^{\dagger}$ is known as the creation operator. Similarly, the operator $\hat{a}$ destroys energy in lumps of $\hbar\omega$ and is known as the annihilation operator. The creation and annihilation operators are central to QFT, and the reader will meet them often in her further studies.

Aside: We often see the creation and annihilation operators written as:

$$
\hat{a}^{\dagger} = \sqrt{\frac{m\omega}{2}}\left(\hat{x} - i\frac{1}{m\omega}\hat{p}\right) = \sqrt{\frac{m\omega}{2}}\left(\hat{x} - \frac{1}{m\omega}\frac{\widehat{\partial}}{\partial x}\right)
$$
$$
\hat{a} = \sqrt{\frac{m\omega}{2}}\left(\hat{x} + i\frac{1}{m\omega}\hat{p}\right) = \sqrt{\frac{m\omega}{2}}\left(\hat{x} + \frac{1}{m\omega}\frac{\widehat{\partial}}{\partial x}\right)
$$

$$(28.41)$$

We have:

$$
\left[\hat{a},\hat{a}^{\dagger}\right] = 1
$$

$$(28.42)$$

EXERCISES

1. Calculate the second energy level of the simple harmonic operator?

2. Verify that the Hamiltonian of the harmonic oscillator can be written as:

$$\widehat{H}_{sho} = \hbar\omega\left(\hat{a}\hat{a}^{\dagger} - \frac{1}{2}\right) \qquad (28.43)$$

3. Using (28.43), calculate $\hat{a}\hat{a}^{\dagger}|n\rangle$?

4. Using the normalized annihilation and creation operators:

$$\hat{a} = \left(\frac{\hbar\omega}{2}\right)^{\frac{1}{2}}\left(x + \frac{\partial}{\partial x}\right)$$

$$\qquad (28.44)$$

$$\hat{a}^{\dagger} = \left(\frac{\hbar\omega}{2}\right)^{\frac{1}{2}}\left(x - \frac{\partial}{\partial x}\right)$$

i) Show $\left[\hat{a}, \hat{a}^{\dagger}\right] = \hbar\omega$

ii) Show $\hat{a}\hat{a}^{\dagger} = \widehat{H} + \frac{1}{2}\hbar\omega$ & $\hat{a}^{\dagger}\hat{a} = \widehat{H} - \frac{1}{2}\hbar\omega$

29

ANGULAR MOMENTUM

Angular momentum appears frequently in Newtonian mechanics. It ought to be no surprise that angular momentum features prominently in quantum mechanics.

The quantum mechanical angular momentum operators can be derived from the Newtonian relations between angular momentum and linear momentum. The angular momentum operators are:

$$\widehat{L_x} = \widehat{y}\widehat{p_z} - \widehat{z}\widehat{p_y} = -i\hbar\left(y\frac{\partial}{\partial z} - z\frac{\partial}{\partial y}\right)$$

$$\widehat{L_y} = \widehat{z}\widehat{p_x} - \widehat{x}\widehat{p_z} = -i\hbar\left(z\frac{\partial}{\partial x} - x\frac{\partial}{\partial z}\right) \qquad (29.1)$$

$$\widehat{L_z} = \widehat{x}\widehat{p_y} - \widehat{y}\widehat{p_x} = -i\hbar\left(x\frac{\partial}{\partial y} - y\frac{\partial}{\partial x}\right)$$

We also have the total angular momentum operator:

$$\widehat{L^2} = \widehat{L_x^2} + \widehat{L_y^2} + \widehat{L_z^2} \qquad (29.2)$$

We most often use the angular momentum operators in polar form. Those polar forms are:

$$\widehat{L_x} = i\hbar\left(\sin\phi\frac{\partial}{\partial\theta} + \cot\theta\cos\phi\frac{\partial}{\partial\phi}\right)$$

$$\widehat{L_y} = -i\hbar\left(\cos\phi\frac{\partial}{\partial\theta} + \cot\theta\sin\phi\frac{\partial}{\partial\phi}\right)$$

$$\widehat{L_z} = -i\hbar\frac{\partial}{\partial\phi} \tag{29.3}$$

$$\widehat{L^2} = -\hbar^2\left(\frac{1}{\tan\theta}\frac{\partial}{\partial\theta} + \frac{\partial^2}{\partial\theta^2} + \frac{1}{\sin^2\theta}\frac{\partial^2}{\partial\phi^2}\right)$$

When we calculate the basis eigenfunctions of these operators, we will be able to write these operators in matrix form. It turns out that there are many different sized matrix representations. For the time being, we give the 3 × 3 representation. We have:

$$\widehat{L^2} = 2\hbar^2\begin{bmatrix} 1 & 0 & 0 \\ 0 & 1 & 0 \\ 0 & 0 & 1 \end{bmatrix} \tag{29.4}$$

And:

$$\widehat{L_z} = \hbar\begin{bmatrix} 1 & 0 & 0 \\ 0 & 0 & 0 \\ 0 & 0 & -1 \end{bmatrix} \tag{29.5}$$

$$\widehat{L_x} = \frac{\hbar}{\sqrt{2}}\begin{bmatrix} 0 & 1 & 0 \\ 1 & 0 & 1 \\ 0 & 1 & 0 \end{bmatrix}, \qquad \widehat{L_y} = \frac{\hbar}{\sqrt{2}}\begin{bmatrix} 0 & -i & 0 \\ i & 0 & -i \\ 0 & i & 0 \end{bmatrix}$$

Remembering that:

$$\left[\widehat{A},\widehat{B}\right]\psi = \widehat{A}\left(\widehat{B}(\psi)\right) - \widehat{B}\left(\widehat{A}(\psi)\right) \tag{29.6}$$

The commutators of the angular momentum operators are:

$$\left[\widehat{L_x},\widehat{L_y}\right] = i\hbar\widehat{L_z}$$

$$\left[\widehat{L_y},\widehat{L_z}\right] = i\hbar\widehat{L_x} \tag{29.7}$$

$$\left[\widehat{L_z},\widehat{L_x}\right] = i\hbar\widehat{L_y}$$

And:

$$\left[\widehat{L^2},L_x\right]=\left[\widehat{L^2},L_y\right]=\left[\widehat{L^2},L_z\right]=0 \qquad (29.8)$$

We see that the total angular momentum operator, $\widehat{L^2}$, commutes with all other angular momentum operators. Since $\widehat{L^2}$ is the sum of the single component angular momentum operators, it ought to be no surprise that the eigenfunctions of all three single component angular momentum operators are within the set of eigenfunctions of $\widehat{L^2}$.

29.1   ANGULAR MOMENTUM IN A CENTRAL FORCE POTENTIAL

The potential energy of a time independent spherical well (a central force) such as the electromagnetic potential of an atomic nucleus varies with only distance from the center (the nucleus). The central force potential is thus of the form $V(r)$. The 3-dimensional Hamiltonian (energy operator) is therefore:

$$\widehat{H} = -\frac{\hbar^2}{2m}\nabla^2 + V(r) \qquad (29.9)$$

After some lengthy algebraic manipulation, this is written as:

$$\widehat{H} = -\frac{\hbar^2}{2m}\left(\frac{\partial^2}{\partial r^2} + \frac{2}{r}\frac{\partial}{\partial r}\right) + \frac{\widehat{L^2}}{2mr^2} + V(r) \qquad (29.10)$$

We see that, disregarding the $\widehat{L^2}$ operator, this Hamiltonian is a function of only r. The four angular momentum operators $\left\{\widehat{L_x},\widehat{L_y},\widehat{L_z},\widehat{L^2}\right\}$ are functions of only $\{\phi, \theta\}$. This means that this Hamiltonian will commute with all four angular momentum operators. The reader is reminded of Ehrenfest's theorem which says that any operator which commutes with the Hamiltonian represents a dynamic variable that is conserved and, in any mechanics, angular momentum is a conserved quantity.

Ehrenfest's theorem says that the expectation value of any time-independent operator is a constant if the time-independent operator commutes with the Hamiltonian. Ehrenfest's theorem is the basis of conservation laws in quantum mechanics. Because the four angular momentum operators commute with the Hamiltonian of a time independent spherical potential, the angular momenta associated with each operator is separately conserved in a time independent spherical potential.

29.2 EIGENVALUE EQUATIONS

Because the Hamiltonian (energy operator), $\widehat{H}$, the total momentum operator, $\widehat{L^2}$, and the z-angular momentum operator, $\widehat{L_z}$ all commute with each other, there will be simultaneous eigenfunctions of these three operators. We need not have chosen $\widehat{L_z}$ but could have instead have chosen either $\widehat{L_x}$ or $\widehat{L_y}$; we chose $\widehat{L_z}$ because it is in a more simple format than either of the other two possible operators. We denote the eigenfunctions of $\{\widehat{H},\widehat{L^2},\widehat{L_z}\}$ as $u(r,\theta,\phi)$. If we were using Dirac notation, we would denote the eigenvectors as $|r, \theta, \phi\rangle$. We anticipate that the solutions of the energy operator (Hamiltonian) eigenvalue equation (Schrödinger's equation) will be separable into radial part and angular part and thus of the form:

$$u(r,\theta,\phi) = R(r)Y(\theta,\phi) \qquad (29.11)$$

The two corresponding eigenvalue equations are:

$$\widehat{H}u(r,\theta,\phi) = E.u(r,\theta,\phi)$$
$$\widehat{L^2}Y(\theta,\phi) = \lambda\hbar^2.Y(\theta,\phi) \qquad (29.12)$$

Note that the angular momentum operators are purely angular, and so we do not need the radial part of the eigenfunctions.

29.3 EIGENVALUES OF THE $\widehat{L_z}$ OPERATOR

The operator, $\widehat{L_z} = -i\hbar\dfrac{\partial}{\partial\phi}$ depends upon only the ϕ angular variable, and so we take it that the $Y(\theta, \phi)$ part of the eigenfunctions will be separable:

$$Y(\theta,\phi) = \Theta(\theta)\Phi(\phi) \qquad (29.13)$$

The third eigenvalue equation is thus:

$$\widehat{L_z}\Phi(\phi) = m\hbar\Phi(\phi)$$

$$-i\hbar\frac{\partial\Phi}{\partial\phi} = m\hbar\Phi \qquad (29.14)$$

With solution:

$$\Phi = N_{L_z}e^{im\phi} \qquad (29.15)$$

Wherein N_{L_z} is a normalization constant.

There is a boundary condition upon these eigenfunctions that derives from the simple fact that $\phi \equiv \phi + 2\pi$. This boundary condition means that we require:

$$N_{L_z}e^{im\phi} = N_{L_z}e^{im(\phi+2\pi)}$$

$$= N_{L_z}e^{im(\phi)}e^{im(2\pi)} \qquad (29.16)$$

Which is:

$$e^{im(2\pi)} = 1 \qquad (29.17)$$

This means that the number m must be a (real) integer, and so we have it that the eigenvalues of the $\widehat{L_z}$ operator are:

$$m\hbar = 0, \ \pm\hbar, \ \pm2\hbar, \ \pm3\hbar,... \qquad (29.18)$$

To know the eigenfunctions associated with these eigenvalues, we need to calculate the normalization constant. We have:

$$\int_0^{2\pi} d\phi \ \Phi^*\Phi = N_{L_z}^2 \int_0^{2\pi} d\phi \ (e^{im\phi})^* e^{im\phi}$$

$$= N_{L_z}^2 \int_0^{2\pi} d\phi \qquad (29.19)$$

$$= 2\pi N_{L_z}^2$$

We require that this be equal to unity:

$$2\pi N_{L_z}^2 = 1$$

$$N_{L_z} = \frac{1}{\sqrt{2\pi}} \qquad (29.20)$$

The eigenfunctions of the $\widehat{L_z}$ angular momentum operator are thus:

$$\Phi = \frac{1}{\sqrt{2\pi}} e^{im\phi} \tag{29.21}$$

29.4 EIGENVALUES OF THE $\widehat{L^2}$ OPERATOR

We now seek the eigenvalues and eigenfunctions of the $\widehat{L^2}$ operator. We have:

$$\widehat{L^2}Y(\theta,\phi) = \lambda\hbar^2.\Theta(\theta)\Phi(\phi)$$
$$\widehat{L^2}\Theta(\theta)\frac{1}{\sqrt{2\pi}}e^{im\phi} = \lambda\hbar^2.\Theta(\theta)\frac{1}{\sqrt{2\pi}}e^{im\phi} \tag{29.22}$$

This is:

$$-\hbar^2\left(\frac{1}{\tan\theta}\frac{\partial}{\partial\theta} + \frac{\partial^2}{\partial\theta^2} + \frac{1}{\sin^2\theta}\frac{\partial^2}{\partial\phi^2}\right)\left(\Theta(\theta)\frac{1}{\sqrt{2\pi}}e^{im\phi}\right)$$
$$= \lambda\hbar^2.\Theta(\theta)\frac{1}{\sqrt{2\pi}}e^{im\phi} \tag{29.23}$$

Differentiating with respect to ϕ and rewriting the first part of the operator leads to:

$$\left(\frac{1}{\sin\theta}\frac{\partial\left(\sin\theta\frac{\partial}{\partial\theta}\right)}{\partial\theta} - \frac{m^2}{\sin^2\theta}\right)\Theta(\theta) = -\lambda\Theta(\theta) \tag{29.24}$$

Substituting $\cos\theta = v$ leads to:

$$\frac{\partial\left((1-v^2)\frac{\partial P(v)}{\partial v}\right)}{\partial v} + \left(\lambda - \frac{m^2}{1-v^2}\right)P(v) = 0 \tag{29.25}$$

This equation is known as the associated Legendre equation. We need to solve this for $P(v)$.

Aside: The Legendre equation is this associated Legendre equation with $m = 0$:

$$\frac{\partial\left((1-v^2)\frac{\partial P(v)}{\partial v}\right)}{\partial v} + \lambda P(v) = 0 \tag{29.26}$$

When $v = \pm 1$, we have division by zero within the associated Legendre equation. However, a series solution exists; it is:

$$P(v) = (1 - v^2)^{\frac{|m|}{2}} \sum_0^\infty a_r v^r \qquad (29.27)$$

Such that the $(r + 2)^{\text{th}}$ element of the series is given by:

$$a_{r+2} = a_r \left(\frac{(r + |m|)(r + |m| + 1) - \lambda}{(r + 1)(r + 2)} \right) \qquad (29.28)$$

This series solution will not lead to division by zero provided the series terminates. We can achieve such termination of the series by choosing:

$$\lambda = (r + |m|)(r + |m| + 1) \equiv l(l + 1) \qquad (29.29)$$

Within this, we have, following convention, taken:

$$l = r + |m| \qquad (29.30)$$

Remembering that r is just an integer counting the terms of the truncated series, this implies:

$$l \geq |m|$$
$$l - |m| = 0, 1, 2, 3, \ldots \qquad (29.31)$$

Clearly, l must be an integer because m is an integer. We now have it that the eigenvalues of the $\widehat{L^2}$ operator are:

$$l(l + 1)\hbar^2 \qquad : \qquad l = 0, 1, 2, 3, \ldots \qquad (29.32)$$

These values correspond to a total angular momentum of $\sqrt{l(l + 1)\hbar^2}$.

The reader might recall that the electrons surrounding an atomic nucleus are arranged into shells respectively referred to by spectroscopists as $\{s, p, d, f, g, h, \ldots\}$. These shells correspond to the values of the total momentum, L_{Total}, as $\{0, 1, 2, 3, 4, 5, \ldots\}$.

The eigenfunctions, $P(v)$, are eigenfunctions of both the $\widehat{L^2}$ and $\widehat{L_z}$ operators. They are separately denoted as $P_l^{|m|}(v)$ and are called the associated Legendre functions. The $|m|$ is justified because only

m^2 appears in the equation that defines them, (29.25). The eigenfunctions are normalized such that:

$$\int_0^\infty d\theta \, \left|\Theta_l^{|m|}\right|^2 \sin\theta = 1 \tag{29.33}$$

29.5 SPHERICAL HARMONICS

The complete angular momentum eigenfunctions are:

$$Y_l^m(\theta,\phi) = \left(\frac{2l+1}{4\pi}\frac{(l+|m|)!}{(l-|m|)!}\right)^{\frac{1}{2}}\left(\frac{-m}{|m|}\right)^{|m|} e^{im\phi} P_l^{|m|}(\cos\theta) \tag{29.34}$$

These are known as the spherical harmonics. They are, or course, the basis of a linear space. That is, the spherical harmonics are complete and mutually orthogonal as defined by the overlap integral:

$$\int_0^\pi d\theta \int_0^{2\pi} d\phi \, \left(Y_j^k\right)^* Y_l^m \sin\theta = \delta_{jl}\delta_{km} \tag{29.34}$$

Wherein the $\delta_{jl}\delta_{km}$ are Kronecker deltas.

The first few normalized spherical harmonics are:

$$Y_0^0 = \left(\frac{1}{4\pi}\right)^{\frac{1}{2}}$$

$$Y_1^1 = -\left(\frac{3}{8\pi}\right)^{\frac{1}{2}} e^{i\phi}\sin\theta$$

$$Y_1^0 = \left(\frac{3}{4\pi}\right)^{\frac{1}{2}}\cos\theta \tag{29.36}$$

$$Y_1^{-1} = \left(\frac{3}{8\pi}\right)^{\frac{1}{2}} e^{-i\phi}\sin\theta$$

$$Y_2^2 = \left(\frac{15}{32\pi}\right)^{\frac{1}{2}} e^{2i\phi}\sin^2\theta$$

In general, the parity of the spherical harmonic depends upon the value of l is given by:

$$(-1)^{l} \qquad (29.37)$$

29.6 ORBITAL ANGULAR MOMENTUM AND ELECTRON ORBITS

Although we did not explicitly say so, the angular momentum with which we have dealt above is known as orbital angular momentum. The reader might wonder what other kind of angular momentum there could possibly be. It was one of the great surprises of quantum mechanics that there is another type of angular momentum which we call intrinsic spin, or just spin. No one really knows what intrinsic spin actually is; it has no counterpart within Newtonian mechanics; it is a purely quantum mechanical phenomenon. We will deal with intrinsic spin shortly. Meanwhile, back to orbital momentum.

The orbital angular momentum vector is described by two eigenvalues (quantum numbers) often denoted by $|l,m\rangle$. Of these, $\sqrt{l(l+1)}\hbar^{2}$ is the total angular momentum. Of these, $m\hbar$ is the z-component of the angular momentum vector. We have above, (29.31), that $l \geq m$. It is not surprising that the total angular momentum is greater than or equal to the z-component of angular momentum. What is surprising is that, not only is the total orbital angular momentum quantitised as Bohr proposed, but that the z-component of the orbital angular momentum is quantitised. This means that orbits within atoms can have only particular angles with the z-axis.

We normally think of an orbiting particle, like the Earth's moon, as being in an orbital plane that slices through the center of the spherical potential. Atoms are not like planets. The orbits of electrons are stacked in layers of orbital planes only one of which (the middle one) slices through the potential center.

Planetary orbits in a Central Potential

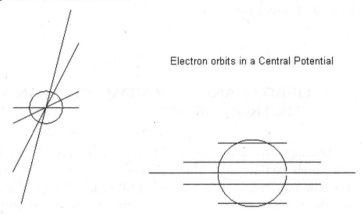

Electron orbits in a Central Potential

The different planes correspond to the different values of the z-component of orbital angular momentum. Since the maximum value of the z-component of orbital angular momentum is $l\hbar$ and this is less than the total orbital angular momentum, $l\hbar$ we see that the angular momentum vector can never be purely in the z-direction – we cannot have an electron orbit of zero radius.

What we have done with the z-component of orbital angular momentum, $\widehat{L_z}$, can be equally well done with either $\widehat{L_x}$ or $\widehat{L_y}$. The mathematics is more complicated, but the results are the same. We chose $\widehat{L_z}$ for pedagogic ease.

SUMMARY

In quantum mechanics, orbital angular momentum is denoted by two quantum numbers, $|l,m\rangle$ which correspond to the total angular momentum $\sqrt{l(l+1)\hbar^2}$ and the z-component of orbital angular momentum, $m\hbar$. We have $l \geq |m|$; this condition is a statement of the fact that total orbital angular momentum is greater than a single component of orbital angular momentum.

EXERCISES

1. A physical system is described by the normalized orbital angular momentum state vector:

$$|\psi\rangle = \frac{1}{\sqrt{21}}\begin{bmatrix} 1 \\ 4 \\ 2 \end{bmatrix} \qquad (29.38)$$

Why is the $\dfrac{1}{\sqrt{21}}$ factor necessary? What are the basis states? If $\widehat{L_z}$ is measured, what results might we get and with what probabilities?

2. Using the block multiplication properties of matrices, substitute $a + ib = \begin{bmatrix} a & b \\ -b & a \end{bmatrix}$ into the matrix $\begin{bmatrix} 0 & -i & 0 \\ i & 0 & -i \\ 0 & i & 0 \end{bmatrix}$ to show that it is a symmetric matrix?

3. Consider the matrix:

$$U = \frac{1}{\sqrt{2}}\begin{bmatrix} -\dfrac{i}{\sqrt{2}} & 1 & \dfrac{i}{\sqrt{2}} \\ 1 & 0 & 1 \\ \dfrac{i}{\sqrt{2}} & 1 & -\dfrac{i}{\sqrt{2}} \end{bmatrix} \qquad (29.39)$$

Is this matrix unitary? If this matrix is unitary, use it to do a unitary transformation upon the matrix:

$$\widehat{L_y} = \frac{\hbar}{\sqrt{2}}\begin{bmatrix} 0 & -i & 0 \\ i & 0 & -i \\ 0 & i & 0 \end{bmatrix} \qquad (29.40)$$

to transform it into the matrix:

$$\widehat{L_z} = \hbar\begin{bmatrix} 1 & 0 & 0 \\ 0 & 0 & 0 \\ 0 & 0 & -1 \end{bmatrix} \qquad (29.41)$$

CHAPTER 30

THE STERN-GERLACH EXPERIMENT

In 1922, in Frankfurt Walther Gerlach (1889–1979) and Otto Stern (1889–1969) conducted an experiment[1] to measure the orbital angular momentum of electrons in a silver atom. Unexpectedly, Gerlach and Stern failed to measure any orbital angular momentum of the electrons but discovered the intrinsic spin angular momentum of the electron. The experiment is known as the Stern-Gerlach experiment.

Although we now say that the Stern-Gerlach experiment discovered intrinsic spin, and it did so, this was not realized at the time and it was several years before Uhlenbeck and Goudsmit hypothesized the existence of intrinsic spin[2]. The experiment was repeated using hydrogen atoms in 1927 by T. E. Phipps and J. B. Taylor[3].

Any electrically charged body that has angular momentum will also have a magnetic dipole moment associated with that angular momentum. The Stern and Gerlach passed a beam of silver atoms through an inhomogeneous magnetic field intending that the magnetic dipole of the electrons in orbit around the silver atom nucleus

---

[1.] Gerlach, W. Stern, O. (1922) Das magnetische moment des silberstoms : Zeitschrift für Physik 9: 353–355.
[2.] S.Goudsmit & G.E.Uhlenbeck Physica 6 (1926) 273.
[3.] Phipps, T.E.: Taylor, J.B. (1927) The magnetic moment of the hydrogen atom. Physical Review 29 (2) 309–320.

would react with this inhomogeneous magnetic field in such a way that the orbital angular momentum of the electrons in the silver atoms could be deduced. This was an experiment looking for a quantitised spectrum of orbital angular momentum eigenvalues. Silver atoms are electrically neutral, and so only the effects of orbiting electrons would be seen.

The magnetic dipole moment of an orbiting electrically charged particle (electrons were thought of in this way in the 1920s) is given by:

$$\vec{\mu} = \frac{q}{2m} = \gamma \vec{L}$$

(30.1)

Wherein q is the electric charge of the particle and m is the mass of the particle. This formula applies to any orbit which conserves angular momentum. It is applicable in quantum mechanics to orbital angular momentum. If the magnetic dipole is in an inhomogeneous magnetic field, the dipole will feel a force displacing it given by:

$$\vec{F} = \vec{\mu} \cdot \nabla \vec{B}$$

(30.2)

The apparatus of the Stern-Gerlach experiment was arranged such that only the z-component of this force was non-zero, and so the force on the dipole in the Stern-Gerlach experiment is given by:

$$F_z = \mu_z \frac{\partial B_z}{\partial z}$$

(30.3)

And we have:

$$F_z = \gamma L_z \frac{\partial B_z}{\partial z}$$

(30.4)

We now incorporate the quantum mechanical view that orbital angular momentum is given by $L_z = m_l \hbar$ where, for a given integer value of l, m_l are the allowed values of the z-component of angular momentum (see previous chapter). We get:

$$F_z = \gamma m_l \hbar \frac{\partial B_z}{\partial z}$$

(30.5)

This says that there will be $(2l + 1)$ different values of force acting on the magnetic dipoles (silver atoms), one for each possible value of

m_l, and so the magnetic dipole (silver atoms) will be deflected by the Stern-Gerlach apparatus into $(2l + 1)$ different beams. Note that the beam associated with $m_l = 0$ will be wholly undeflected.

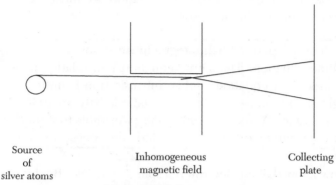

Source
of
silver atoms

Inhomogeneous
magnetic field

Collecting
plate

The Stern-Gerlach Apparatus

We can determine the values of the number of different values of orbital angular momentum from counting the number of beams. If, instead of a number of beams, we get just a smudge, then the value of orbital angular momentum is not discreet, as quantum mechanics says it should be, but continuous as classical physics would predict.

When the Stern-Gerlach experiment was first done, it produced only two beams of silver atoms (as shown above). Two does not equal $(2l + 1)$ for any integer value of l. We now know that the orbital angular momentum of a silver atom is zero; this would correspond to only one beam, but the experiment did not produce only one beam. That there are two beams means there are two, and only two, possible values of magnetic dipole (angular momentum) in the silver atom.

The deflection of silver atoms into beams in the Stern-Gerlach experiment is due to the intrinsic spin angular momentum of the single (unpaired) electron in a silver atom. If intrinsic spin is to satisfy similar rules as orbital angular momentum, then there should be a $(2s + 1)$ rule. Two beams mean that $m_s = \pm\dfrac{\hbar}{2}$. Intrinsic spin angular momentum comes in half integer values.

Intrinsic spin is thought of in quantum mechanics as a type of angular momentum that is different from orbital angular momentum.

There is no Newtonian counterpart to intrinsic spin, and, therefore, no Newtonian variable that corresponds to the spin operator. There is nothing like intrinsic spin predicted by the Schrödinger equation. If we wish to use the Schrödinger equation, we must arbitrarily add in the concept of spin (see later).

Aside: This property of half integral intrinsic spin does emerge from the Dirac equation. The Dirac equation is a relativistic version of the Schrödinger equation, and so intrinsic spin is unthinkingly, and wrongly, seen as necessitated by special relativity. In 1934, Wolfgang Pauli and Victor Weisskopf (1908–2002) were able to show that a relativistic quantum theory need not include the concept of spin at all[4].

We now know that an electron is a spin $\frac{1}{2}$ particle. By this, we mean that, along any axis, the eigenvalues of the intrinsic spin operator (to be introduced in the next chapter) are only two, $\pm\frac{\hbar}{2}$. It does not matter in what direction we orientate the axis we choose to measure against, we will still get only a half integral value of intrinsic spin for an electron. The amount of intrinsic spin of a particle, $\left\{0,\frac{1}{2},1,\frac{3}{2},...\right\}$, be it electron, photon, quark, Higgs boson, or whatever is a fixed property of that particle; it cannot be increased or decreased. Spin $\frac{1}{2}$ particles have two possible values of intrinsic spin, $\pm\frac{1}{2}\hbar$. Spin 1 particles (like the W and Z bosons) have three possible values of intrinsic spin $\{-1, 0, +1\}$.

30.1 THE ZEEMAN EFFECT

The spectral lines emitted or absorbed by an atom change slightly when the atom is within a magnetic field. This is due to the change in energy levels within the atom brought about by the

---

[4.] W. Pauli & V.F. Weisskopf, Helv. Phys. Acta 7,709 (1934).

interaction of the magnetic field and the magnetic dipoles associated with the angular momenta of the particles that comprise the atom:

$$E_{mag} = -\vec{\mu} \cdot \vec{B}$$
$$= -\gamma L_z B \qquad (30.6)$$
$$= -\gamma m_l \hbar B$$

Wherein we have assumed that the magnetic field is in the z-direction. m_l is called the magnetic quantum number, and γ is the magneto-gyric ratio. This effect is very useful in exploring the structure of atoms.

Because of this energy change, we need to modify the energy operator in the time independent Schrödinger equation by adding E_{mag}. We have:

$$\widehat{H_0}\psi = E_n\psi \rightarrow \left(\widehat{H_0} - \gamma B\widehat{L_z}\right)\psi = E_n\psi \qquad (30.7)$$

For an electron, the extra energy is:

$$E_{Mag} = \frac{e\hbar}{2m}Bm_l \qquad (30.8)$$

The quantity $\dfrac{e\hbar}{2m}$ is called the Bohr magneton and is often denoted by μ_B.

Aside: There are many Stern Gerlach experiments that have never been done but are reported in books on quantum mechanics. The essence of these experiments is a series of three Stern Gerlach magnets. The first Stern Gerlach magnet is oriented with the z-axis vertical and thus produces two beams of silver atoms. One of the beams of silver atoms is then blocked. The second Stern Gerlach magnet receives the remaining beam of silver atoms. The second Stern Gerlach magnet is oriented with the z-axis at $45°$ to the first Stern Gerlach magnet and produces two beams of silver atoms. One of the beams of silver atoms is then blocked. The third Stern Gerlach magnet receives the remaining beam of silver atoms. The third Stern Gerlach magnet is oriented with the z-axis vertical to match the first Stern Gerlach magnet. We might expect that, since one beam of silver atoms was blocked at the exit of the first Stern Gerlach magnet, the third Stern Gerlach magnet would produce only one beam of silver atoms. It does not produce only one beam of silver atoms; it produces two such beams.

The eigenstate of the silver atoms in the single beam exiting the first Stern Gerlach magnet does not persist through the 45° measurement of the second Stern Gerlach magnet.

The experiment has never been done because it is extremely difficult to isolate the apparatus from stray magnetic fields that would precess the dipole of the silver atoms between magnets. It is presented as an example of the nature of the superposition wavefunction in quantum mechanics.

31

INTRINSIC SPIN

The energy levels of electrons within atoms correspond to the spectral lines of those atoms. It was noticed in the first part of the 20th century that spectral lines come in pairs. A particular clear example is sodium it which the $3p \rightarrow 3s$ transition corresponds to the spectral lines at 5896 Angstroms and 5890 Angstroms. It was Uhlenbeck and Goudsmit who first suggested that such doubling of spectral lines was caused by a kind of "internal angular momentum"[1]. We now refer to that "internal angular momentum" as intrinsic spin.

31.1 DIFFERENT SIZES OF ANGULAR MOMENTUM OPERATORS

The reader will recall that once we have the eigenfunctions of an operator, we can calculate the matrix form of that operator from the formula (8.38):

$$M_{RC} = \int_{-\infty}^{\infty} dx \; \phi_R{}^{\circ} \widehat{A}(\phi_C) \qquad (31.1)$$

Looking at the partial list of the eigenfunctions of the orbital angular momentum z-component (spherical harmonics), (29.36), we see that when $l = 0$, we have only one eigenfunction. This means that for $l = 0$,

---

[1] S.Goudsmit and G.Uhlenbeck. Naturwiss, 13, 953 (1925) & Nature 117, 264 (1926).

the matrix form of the $\widehat{L_z}$ operator will be a 1×1 matrix. When $l = 1$, we have three eigenfunctions which are:

$$Y_1^1 = -\left(\frac{3}{8\pi}\right)^{\frac{1}{2}} e^{i\phi} \sin\theta$$

$$Y_1^0 = \left(\frac{3}{4\pi}\right)^{\frac{1}{2}} \cos\theta \qquad (31.2)$$

$$Y_1^{-1} = \left(\frac{3}{8\pi}\right)^{\frac{1}{2}} e^{-i\phi} \sin\theta$$

Using the formula, (31.1), will give us a 3×3 matrix form of the $\widehat{L_z}$ operator. We see these matrices listed in (29.5). There are five $l = 2$ eigenfunctions $\left\{Y_2^{-2}, Y_2^{-1}, Y_2^0, Y_2^1, Y_2^2\right\}$, and these will give a 5×5 $\widehat{L_z}$ matrix operator. With a little thought, the reader will see that, as l increases, the size of the operator matrix will increase but it will always be of an odd size. The reader might wonder if there are any even sized $n \times n$ operator matrices that satisfy the same commutation relations, (29.7) & (29.8), as the $\left\{\widehat{L_x}, \widehat{L_y}, \widehat{L_z}, \widehat{L^2}\right\}$ operators; perhaps a 2×2 matrix or a 4×4 matrix. If there are even sized matrices satisfying the same commutation relations as angular momentum, then these even sized matrices will presumably be something to do with angular momentum – same commutation relations!

Well, there are even sized matrices that have the same commutation relations as the $\left\{\widehat{L_x}, \widehat{L_y}, \widehat{L_z}, \widehat{L^2}\right\}$ operators. The 2×2 matrices are:

$$\widehat{L_x} = \frac{\hbar}{2}\begin{bmatrix} 0 & 1 \\ 1 & 0 \end{bmatrix}, \quad \widehat{L_y} = \frac{\hbar}{2}\begin{bmatrix} 0 & -i \\ i & 0 \end{bmatrix}$$

$$\widehat{L_z} = \frac{\hbar}{2}\begin{bmatrix} 1 & 0 \\ 0 & -1 \end{bmatrix}, \quad \widehat{L^2} = \frac{3\hbar^2}{4}\begin{bmatrix} 1 & 0 \\ 0 & 1 \end{bmatrix} \qquad (31.3)$$

These are universally[2] known as the Pauli matrices. They were introduced *ad hoc* by Wolfgang Pauli in 1926 in an attempt to correct the erroneous prediction of the Schrödinger equation that the magnetic moment of the ground state hydrogen atom was zero.

[2.] That is universally on Earth.

31.2 THE ANGULAR MOMENTUM SPECTRUM FROM THE COMMUTATOR RELATIONS

The set of eigenvalues of an operator is called the spectrum of that operator. We have a set of commutation relations (for angular momentum) that can be written as many different sized square matrices. These different sized matrices correspond to different values of the eigenvalues, l, of the angular momentum operators $\widehat{L^2}$. If we can, which we shortly will, discover all the different sizes of matrix that have the same commutation relations as the angular momentum operators, then we will know the whole set of possible values (eigenvalues) of $\{l, m\}$.

We begin be defining two new operators as a sum of two angular momentum operators.

$$\widehat{L_+} = \widehat{L_x} + i\widehat{L_y}$$
$$\widehat{L_-} = \widehat{L_x} - i\widehat{L_y} \tag{31.4}$$

These are not Hermitian operators, but that is unimportant. They are also known as ladder operators. The commutation relations of these new operators are:

$$\left[\widehat{L_z}, \widehat{L_+}\right] = +\hbar\widehat{L_+} \qquad : \qquad \left[\widehat{L_z}, \widehat{L_-}\right] = -\hbar\widehat{L_-}$$
$$\left[\widehat{L^2}, \widehat{L_+}\right] = 0 \qquad : \qquad \left[\widehat{L^2}, \widehat{L_-}\right] = 0 \tag{31.5}$$

And:

$$\widehat{L_+}\left(\widehat{L_-}(\psi)\right) = \widehat{L^2}(\psi) - \widehat{L_z}\left(\widehat{L_z}(\psi)\right) + \hbar\widehat{L_z}(\psi)$$
$$\widehat{L_-}\left(\widehat{L_+}(\psi)\right) = \widehat{L^2}(\psi) - \widehat{L_z}\left(\widehat{L_z}(\psi)\right) - \hbar\widehat{L_z}(\psi) \tag{31.6}$$

Because $\widehat{L_+}$ commutes with $\widehat{L^2}$, we have:

$$\widehat{L_+}\left(\widehat{L^2}(\psi)\right) = \widehat{L^2}\left(\widehat{L_+}(\psi)\right) \tag{31.7}$$

But:

$$\widehat{L_+}\left(\widehat{L^2}(\psi)\right) = \widehat{L_+}\left(\lambda(\psi)\right) = \lambda\widehat{L_+}(\psi) \tag{31.8}$$

So:

$$\widehat{L^2}\left(\widehat{L_+}(\psi)\right) = \lambda\widehat{L_+}(\psi) \tag{31.9}$$

In other words, $\widehat{L_+}(\psi)$ is an eigenvector of $\widehat{L^2}$. Similarly, $\widehat{L_-}(\psi)$ is an eigenvector of $\widehat{L^2}$. We have, (31.5):

$$\widehat{L_z}\left(\widehat{L_+}(\psi)\right) - \widehat{L_+}\left(\widehat{L_z}(\psi)\right) = +\hbar\widehat{L_+}(\psi)$$

$$\widehat{L_+}\left(\widehat{L_z}(\psi)\right) = \widehat{L_z}\left(\widehat{L_+}(\psi)\right) - \hbar\widehat{L_+}(\psi)$$

(31.10)

And:

$$\widehat{L_+}\left(\widehat{L_z}(\psi)\right) = \widehat{L_+}(\mu\psi) = \mu\widehat{L_+}(\psi)$$

(31.11)

Putting these together gives:

$$\mu\widehat{L_+}(\psi) = \widehat{L_z}\left(\widehat{L_+}(\psi)\right) - \hbar\widehat{L_+}(\psi)$$

$$\widehat{L_z}\left(\widehat{L_+}(\psi)\right) = \mu\widehat{L_+}(\psi) + \hbar\widehat{L_+}(\psi)$$

$$= (\mu + \hbar)\widehat{L_+}(\psi)$$

(31.12)

We see that the $\widehat{L_+}$ operator acts upon a first eigenvector of the $\widehat{L_z}$ operator to produce a second eigenvector of the $\widehat{L_z}$ operator. The second eigenvector has an eigenvalue that is greater than the eigenvalue of the first eigenvector by a single $\hbar$. Similarly, the $\widehat{L_-}$ operator acts upon a first eigenvector of the $\widehat{L_z}$ operator to produce a second eigenvector of the $\widehat{L_z}$ operator such that the second eigenvector has an eigenvalue that is less than the eigenvalue of the first eigenvector by a single $\hbar$.

The eigenvalues of the $\widehat{L_z}$ operator (the z-component of angular momentum) must be less than the eigenvalues of the total angular momentum operator, and so the $\widehat{L_+}$ operator cannot continue forever producing eigenvectors of the $\widehat{L_z}$ operator with higher eigenvalues but must eventually come to act upon the "top eigenvector" with the highest eigenvalue that is possible. When the $\widehat{L_+}$ operator acts upon the "top eigenvector" of the $\widehat{L_z}$ operator, it must produce an output of zero, $\widehat{L_+}(\psi_{Top}) = 0$. Similarly, the $\widehat{L_-}$ operator will eventually come to act upon the "bottom eigenvector" and will produce zero, $\widehat{L_-}(\psi_{Bottom}) = 0$.

Since $\widehat{L_+}(\psi_{Top}) = 0$:

$$\widehat{L_-}\left(\widehat{L_+}\left(\psi_{Top}\right)\right)=0 \tag{31.13}$$

But, (31.6):

$$\widehat{L_-}\left(\widehat{L_+}\left(\psi\right)\right)=\widehat{L^2}\left(\psi\right)-\widehat{L_z}\left(\widehat{L_z}\left(\psi\right)\right)-\hbar\widehat{L_z}\left(\psi\right) \tag{31.14}$$

So:

$$\widehat{L^2}\left(\psi_{Top}\right)-\widehat{L_z}\left(\widehat{L_z}\left(\psi_{Top}\right)\right)-\hbar\widehat{L_z}\left(\psi_{Top}\right)=0$$

$$\left(\lambda-\mu_{Top}^2-\hbar\mu_{Top}\right)=0 \tag{31.15}$$

$$\lambda=\mu_{Top}\left(\mu_{Top}+\hbar\right)$$

Similarly:

$$\lambda=\mu_{Bottom}\left(\mu_{Bottom}-\hbar\right) \tag{31.16}$$

We know that $\mu_{Top}=-\mu_{Bottom}$. The equations (31.15) and (31.16) together lead to:

$$\mu_{Top}-\mu_{Bottom}=\hbar$$

$$2\mu_{Top}=\hbar \tag{31.17}$$

We known the steps between eigenvalues are multiples of $\hbar$ and so we have:

$$\mu_{Top}=\frac{m\hbar}{2} \tag{31.18}$$

Leading to:

$$l=0,\frac{1}{2},1,\frac{3}{2},... \tag{31.19}$$

31.3 ANGULAR MOMENTUM COMES IN HALF-INTEGRAL UNITS

What we have here is that the commutation relations of the angular momentum operators have compelled the eigenvalues of angular momentum to be half-integral units of $\hbar$. The lowest non-trivial angular momentum is $l=\frac{1}{2}$, $m=\pm\frac{1}{2}$ and this gives the matrices (31.3). There are matrices of sizes that are all even values of $2l+1$.

31.4 REPRESENTATIONS OF COMMUTATION RELATIONS

There are 2 × 2 matrices that have the commutation relations of angular momentum. These are called the 2-dimensional representation of the commutation relations – the 2-dimensional representation of angular momentum corresponding to the $l = \frac{1}{2}, m = \pm\frac{1}{2}$ eigenvalues. There are 4 × 4 matrices that have the commutation relations of angular momentum. These are called the 3-dimensional representation of the commutation relations – the 3-dimensional representation of angular momentum corresponding to the $l = 1$, $m = 0, \pm 1$ eigenvalues. There are 4 × 4 matrices that have the commutation relations of angular momentum. These are called the 4-dimensional representation of the commutation relations – the 4-dimensional representation of angular momentum corresponding to the $l = \frac{3}{2}, m = \pm\frac{1}{2}, \pm\frac{3}{2}$ eigenvalues. And so on…

31.5 WHAT IS INTRINSIC SPIN?

The integral eigenvalues of angular momentum arose from the imposition of the boundary condition $\psi(\phi + 2\pi) = \psi(\phi)$. This boundary condition does not require the existence of half-integral eigenvalues of angular momentum; they are allowed, but not required, by the commutation relations. We see that there can be two types of angular momentum, integral and half-integral. Because the half-integral angular momentum does not arise from the spatial boundary condition $\psi(\phi + 2\pi) = \psi(\phi)$, it is thought of as being intrinsic to the orbiting particle (rather than intrinsic to the spatial orbit). The half integral angular momentum is called intrinsic spin, or just spin.

Intrinsic spin was first suspected in 1925 by Goudsmit and Uhlenbeck[3]. They introduced it in an *ad hoc* manner to explain particular spectral lines. In 1928, Dirac produced the Dirac equation

---

[3.] S.Goudsmit & G.E.Uhlenbeck Physica 6 (1926) 273.

as a way of reconciling special relativity with quantum mechanics. The Dirac equation is used in QFT instead of Schrödinger's equation. The Dirac equation automatically has half-integral spin. This is why intrinsic spin is often said to be a relativistic effect. If the Dirac equation is a correct description of nature, and it seems to be that, then intrinsic spin must exist.

Some atomic particles have integral spin; pions, photons, some mesons, and bosons in general are of this nature. Some atomic particles have half-integral spin; electrons, quarks, Ω^-, and fermions in general are of this nature. Integral spin particles are dealt with in QFT by the Klein-Gorden equation. Half-integral spin particles are dealt with by the Dirac equation. The properties of the two types of particles are quite different. A particularly important difference is that the half-integral spin particles obey the Pauli exclusion principle whereas particles with integral span do not obey the Pauli exclusion principle.

Aside: The Austrian physicist Wolfgang Ernst Pauli (1900–1958) was, in 1930, the first to postulate the existence of the neutrino. In 1940, he proved the spin-statistics theorem (see later), and, in 1945, he was awarded the physics Nobel prize. He is most famous for the Pauli spin matrices and for the Pauli exclusion principle.

Among physicists of the time, it was reputed that Pauli had the ability to break laboratory equipment by simply standing close to it. This reputation became so engrained that it gained its own name; it was known as the Pauli effect, almost as if it were a physical law.

31.6 SPIN OPERATORS AND SPIN EIGENVECTORS

Intrinsic spin has only two possible eigenvectors called respectively, and picturesquely but not accurately, spin up, $+\frac{\hbar}{2}$, and spin down, $-\frac{\hbar}{2}$. If we had known this, perhaps from observation, we would have known that the operator associated with these two eigenvectors would be a Hermitian 2×2 matrix. Realizing that angular momentum has three components would have led to three Hermitian 2×2 matrices, and then to the total spin matrix. Since intrinsic spin is a different kind of angular momentum to the orbital kind, we denote the intrinsic spin operators as $\left\{ \widehat{S^2}, \widehat{S_x}, \widehat{S_y}, \widehat{S_z} \right\}$.

We choose to write the matrix operators with basis vectors that are the simultaneous eigenvectors of the intrinsic spin operators $\left\{ \widehat{S^2}, \widehat{S_z} \right\}$. Because we choose this basis, the $\left\{ \widehat{S^2}, \widehat{S_z} \right\}$ operators will be diagonal matrices with the eigenvalues on the leading diagonal. We therefore have:

$$\widehat{S_z} = \frac{\hbar}{2} \begin{bmatrix} 1 & 0 \\ 0 & -1 \end{bmatrix} \qquad : \qquad \widehat{S^2} = \frac{3\hbar^2}{4} \begin{bmatrix} 1 & 0 \\ 0 & 1 \end{bmatrix} \qquad (31.20)$$

Now $\left\{ \widehat{S_x}, \widehat{S_y} \right\}$ do not commute with $\widehat{S_z}$. This means that they are not diagonal unless written in a different basis to $\widehat{S_z}$, and to each other. This means that, in the $\widehat{S_z}$ basis, the elements on the leading diagonals of the $\left\{ \widehat{S_x}, \widehat{S_y} \right\}$ operators will be zero. This together with the hermicity property and the necessary commutation relations leads to the two operators:

$$\widehat{S_x} = \frac{\hbar}{2} \begin{bmatrix} 0 & 1 \\ 1 & 0 \end{bmatrix} \qquad : \qquad \widehat{S_y} = \frac{\hbar}{2} \begin{bmatrix} 0 & -i \\ i & 0 \end{bmatrix} \qquad (31.21)$$

It was shown in 1928 by Paul Dirac that these matrices emerge naturally from the Dirac equation. The reader will often see these 2×2 matrices written as $\left\{ \widehat{\sigma_x}, \widehat{\sigma_y}, \widehat{\sigma_z} \right\}$ rather than the more general $\left\{ \widehat{S_x}, \widehat{S_y}, \widehat{S_z} \right\}$.

The eigenvectors are found by solving the eigenvalue equations. The eigenvectors are ordered pairs of complex numbers of the form:

$$\begin{bmatrix} a + ib \\ c + id \end{bmatrix} \qquad (31.22)$$

These pairs of complex numbers (these complex vectors in $\mathbb{C}^2$) are called spinors.

Aside: There seems to be no agreement as to the pronunciation of the word spinor; it is spelt as if the "i" were acute, but it is often pronounced as if it were spelt "spinnor".

We have:

$$\widehat{S_z}\begin{bmatrix} a+ib \\ c+id \end{bmatrix} = \frac{\hbar}{2}\begin{bmatrix} 1 & 0 \\ 0 & -1 \end{bmatrix}\begin{bmatrix} a+ib \\ c+id \end{bmatrix} = \mu\begin{bmatrix} a+ib \\ -c-id \end{bmatrix} \qquad (31.23)$$

Since $\widehat{S_z}$ commutes with $\widehat{S^2}$, they must have the same basis (complex) vectors, and so we have:

$$\widehat{S^2}\begin{bmatrix} a+ib \\ c+id \end{bmatrix} = \frac{3\hbar}{4}\begin{bmatrix} 1 & 0 \\ 0 & 1 \end{bmatrix}\begin{bmatrix} a+ib \\ c+id \end{bmatrix} = \lambda\begin{bmatrix} a+ib \\ -c-id \end{bmatrix} \qquad (31.24)$$

We see that the spinor will be an eigenvector of both the operators $\{\widehat{S_z}, \widehat{S^2}\}$ with eigenvalues $\lambda = \dfrac{3\hbar^2}{4}$, $\mu = \pm\dfrac{\hbar}{2}$ if $\{a = c = 1, b = d = 0\}$, and so we have the eigenvectors and eigenvalues of $\{\widehat{S_z}, \widehat{S^2}\}$:

$$\begin{bmatrix} 1 \\ 0 \end{bmatrix} : \mu = \frac{\hbar}{2} \qquad\qquad \begin{bmatrix} 0 \\ 1 \end{bmatrix} : \mu = -\frac{\hbar}{2} \qquad (31.25)$$

In the same basis, the eigenvectors of $\{\widehat{S_x}, \widehat{S_y}\}$ with $\mu = \pm\dfrac{\hbar}{2}$ are:

$$\widehat{S_x} :: \quad \frac{1}{\sqrt{2}}\begin{bmatrix} 1 \\ 1 \end{bmatrix} : \frac{\hbar}{2} \quad : \quad \frac{1}{\sqrt{2}}\begin{bmatrix} -1 \\ 1 \end{bmatrix} : -\frac{\hbar}{2}$$

$$\widehat{S_y} :: \quad \frac{1}{\sqrt{2}}\begin{bmatrix} 1 \\ i \end{bmatrix} : \frac{\hbar}{2} \quad : \quad \frac{1}{\sqrt{2}}\begin{bmatrix} 1 \\ -i \end{bmatrix} : -\frac{\hbar}{2} \qquad (31.26)$$

Remember that the inner product of two 2-component complex vectors is:

$$\langle\psi|\phi\rangle = \begin{bmatrix} a^* & b^* \end{bmatrix}\begin{bmatrix} c \\ d \end{bmatrix} \qquad (31.27)$$

31.7 UNITARY TRANSFORMATIONS OF SPIN MATRICES

The three matrices, $\{\widehat{S_x}, \widehat{S_y}, \widehat{S_z}\}$ each represent spin in a different direction. With a little thought, the reader will realize that they must be the same matrix written in three different bases which correspond to three re-orientations of the co-ordinate system. We can transform each of the spin matrices into the other spin matrices with a similarity transformation (remember that a similarity transformation is a change of basis) done with a unitary matrix. The unitary matrix that transforms $\widehat{S}_x$ into $\widehat{S}_z$ is:

$$U = \frac{1}{\sqrt{2}}\begin{bmatrix} 1 & 1 \\ 1 & -1 \end{bmatrix} \tag{31.28}$$

Note that we have:

$$U^{\dagger}U = \frac{1}{\sqrt{2}}\begin{bmatrix} 1 & 1 \\ 1 & -1 \end{bmatrix}\frac{1}{\sqrt{2}}\begin{bmatrix} 1 & 1 \\ 1 & -1 \end{bmatrix} = \begin{bmatrix} 1 & 0 \\ 0 & 1 \end{bmatrix} \tag{31.29}$$

We have:

$$U^{\dagger}\widehat{S}_x U = \frac{1}{\sqrt{2}}\begin{bmatrix} 1 & 1 \\ 1 & -1 \end{bmatrix}\frac{\hbar}{2}\begin{bmatrix} 0 & 1 \\ 1 & 0 \end{bmatrix}\frac{1}{\sqrt{2}}\begin{bmatrix} 1 & 1 \\ 1 & -1 \end{bmatrix} = \frac{\hbar}{2}\begin{bmatrix} 1 & 0 \\ 0 & -1 \end{bmatrix} = \widehat{S}_z \tag{31.30}$$

The above is illustrative of a general feature of non-commuting operators in quantum mechanics. If two quantum mechanical operators do not commute, they have different eigenfunctions (eigenvectors), but the eigenfunctions (eigenvectors) differ only in that they are written in a different basis. We can change the eigenfunctions of one of a pair of non-commutating operators into the eigenfunctions of the other of the pair of non-commutating operators by a similarity transformation using a unitary matrix – a unitary transformation.

In fact, the Pauli spin matrices together with the identity matrix form a 4-dimensional basis. We have:

$$\begin{bmatrix} a+d & 0 \\ 0 & a+d \end{bmatrix}\frac{1}{2}\begin{bmatrix} 1 & 0 \\ 0 & 1 \end{bmatrix} + \begin{bmatrix} b+c & 0 \\ 0 & b+c \end{bmatrix}\frac{1}{2}\begin{bmatrix} 0 & 1 \\ 1 & 0 \end{bmatrix}$$
$$+ \begin{bmatrix} ib-ic & \\ & ib-ic \end{bmatrix}\frac{1}{2}\begin{bmatrix} 0 & -i \\ i & 0 \end{bmatrix} + \begin{bmatrix} a-d & 0 \\ 0 & a-d \end{bmatrix}\frac{1}{2}\begin{bmatrix} 1 & 0 \\ 0 & -1 \end{bmatrix} \tag{31.31}$$
$$= \frac{1}{2}\begin{bmatrix} a+d+a-d & b+c+b-c \\ b+c-b+c & a+d-a+d \end{bmatrix} = \begin{bmatrix} a & b \\ c & d \end{bmatrix}$$

EXERCISES

1. Verify the orthonormality of the spinors of $\widehat{S}_z$.

2. What effect does a similarity transformation with the unitary matrix (31.28) have upon $\widehat{S}_y$?

3. What unitary matrix transforms $\widehat{S}_y$ to $\widehat{S}_z$?

4. What are the squares of the Pauli matrices? Are these square roots of unity?

5. Show that the Pauli matrices, $\widehat{\sigma}_i$ relate to each other as:

$$\widehat{\sigma}_x\widehat{\sigma}_y = -\widehat{\sigma}_y\widehat{\sigma}_x = i\widehat{\sigma}_z \qquad (31.32)$$

6. We have:

$$i\sigma_z = \begin{bmatrix} i & 0 \\ 0 & -i \end{bmatrix} \qquad (31.33)$$

Use:

$$a + ib = \begin{bmatrix} a & b \\ -b & a \end{bmatrix} \qquad (31.34)$$

And the block multiplication properties of matrices to write $i\sigma z$ as a 4×4 matrix. Is this 4×4 matrix symmetric? Are the eigenvalues of symmetric matrices always real?

32

A REVISION OF THE STRUCTURE OF QUANTUM MECHANICS

Intrinsic spin is a very simple quantum mechanical system comprised of only two eigenvectors. Because it is so simple, we can use it to demonstrate the structure of quantum mechanics. This chapter will serve as a revision of what as gone before. We present this chapter to "ram home" some of the previous chapters.

We will consider only intrinsic spin, and we will ignore all other aspects of a physical system. As is conventional, we will denote the spin up eigenstate as $|\uparrow\rangle$; this eigenstate has eigenvalue $\dfrac{\hbar}{2}$. We denote the spin down eigenstate as $|\downarrow\rangle$; ; this eigenstate has eigenvalue $-\dfrac{\hbar}{2}$. We work with $\widehat{S_z}$.

1. The general quantum mechanical state, $|\psi\rangle$, is a superposition (linear sum) of the basis eigenstates. As far as $\widehat{S_z}$ is concerned, this is:

$$|\psi\rangle = c_1|\uparrow\rangle + c_2|\downarrow\rangle = (a + ib)\begin{bmatrix} 1 \\ 0 \end{bmatrix} + (c + id)\begin{bmatrix} 0 \\ 1 \end{bmatrix} \qquad (32.1)$$

We reiterate that the coefficients (amplitudes), c_i, are complex, and we reiterate that for most dynamic variables, the vectors (eigenstates) have complex components, $\in \mathbb{C}^n$. We take it that the coefficients are normalized.

2. The probability that, upon observation, a physical system, say an electron, will be found to be in a particular eigenstate, ψ_i, is given by the modulus of the coefficient of that eigenstate. The probability that the electron has the $\widehat{S}_z$ component of spin up, $|\uparrow\rangle$, that is with eigenvalue $\dfrac{\hbar}{2}$, is given by $|c_1|^2 = (a + ib)(a - ib)$. The probability that the electron has the $\widehat{}$ component of spin down, $|\downarrow\rangle$, that is with eigenvalue $-\dfrac{\hbar}{2}$, is given by $|c_2|^2 = (c + id)(c - id)$.

3. Since we assumed the state $|\psi\rangle$ was normalized, we have:

$$|c_1|^2 + |c_2|^2 = 1 \tag{32.2}$$

As we expect, total probability is equal to unity.

4. The expectation value (roughly average value) of the z-component of spin is:

$$
\begin{aligned}
\left\langle \widehat{S}_z \right\rangle = \left\langle \psi \middle| \widehat{S}_z \middle| \psi \right\rangle &= \frac{\hbar}{2} \begin{bmatrix} c_1^\circ & c_2^\circ \end{bmatrix} \begin{bmatrix} 1 & 0 \\ 0 & -1 \end{bmatrix} \begin{bmatrix} c_1 \\ c_2 \end{bmatrix} \\
&= \frac{\hbar}{2} \left(|c_1|^2 - |c_2|^2 \right) \\
&= \frac{\hbar}{2} |c_1|^2 + \left(-\frac{\hbar}{2} \right) |c_2|^2
\end{aligned}
\tag{32.3}
$$

Since the z-component of spin operator, $\widehat{S}_z$, and the total spin operator, $\widehat{S^2}$, commute, the superposition of eigenstates of these operators is written in the same basis – they have the same eigenstates – and we also have the expectation value of the total spin operator as:

$$
\begin{aligned}
\left\langle \widehat{S^2} \right\rangle &= \frac{3\hbar^2}{4} \begin{bmatrix} c_1^\circ & c_2^\circ \end{bmatrix} \begin{bmatrix} 1 & 0 \\ 0 & 1 \end{bmatrix} \begin{bmatrix} c_1 \\ c_2 \end{bmatrix} \\
&= \frac{3\hbar^2}{4} \left(|c_1|^2 + |c_2|^2 \right) \\
&= \frac{3\hbar^2}{4}
\end{aligned}
\tag{32.4}
$$

The direction we choose to be the z-axis is arbitrary. It would not matter what direction we chose as the z-axis, the only possible eigenstates would be the spin-up, $|\uparrow\rangle$, state and the spin down state, $|\downarrow\rangle$, with eigenvalues $\dfrac{\hbar}{2}, -\dfrac{\hbar}{2}$ respectively.

EXERCISES

1. Using the 2×2 matrix form of the complex numbers and the block multiplication property of matrices, (This allows you to double the size of a matrix by replacing a complex number element by the 2×2 complex number matrix.) and the matrices and expressions above (31.3) and (31.4), write $\left\{\widehat{L_+}, \widehat{L_-}\right\}$ as 4×4 matrices. Are these symmetric or anti-symmetric matrices?

2. What are the eigenvalues of the matrix $\dfrac{\hbar}{2}\begin{bmatrix} 0 & 1 \\ 1 & 0 \end{bmatrix}$?

3. What are the eigenvectors of the matrix $\dfrac{\hbar}{2}\begin{bmatrix} 0 & 1 \\ 1 & 0 \end{bmatrix}$?

4. If a spin $\dfrac{1}{2}$ system is in the state $|\phi\rangle = \dfrac{1}{\sqrt{6}}\begin{bmatrix} \sqrt{5} \\ 1 \end{bmatrix}$, what is the probability that the system will be observed to be in the spin $\begin{bmatrix} 1 \\ 0 \end{bmatrix} = Spin - up$ state?

5. Normalize the state $\begin{bmatrix} 6 \\ 3 \end{bmatrix}$

33

A STRANGE ASPECT OF INTRINSIC SPIN

Suppose that we have just measured the spin of an electron to be in a particular direction. We now know with certainty that the spin is in that direction. We have:

$$\widehat{S_z}\begin{bmatrix} a+ib \\ c+id \end{bmatrix} = \frac{\hbar}{2}\begin{bmatrix} a+ib \\ c+id \end{bmatrix} \tag{33.1}$$

We now choose to re-orient our axes by rotating them through the angle θ. We now have:

$$\widehat{S_{z\theta}} = \widehat{S_z}\cos\theta + \widehat{S_x}\sin\theta$$

$$= \frac{\hbar}{2}\begin{bmatrix} \cos\theta & 0 \\ 0 & -\cos\theta \end{bmatrix} + \frac{\hbar}{2}\begin{bmatrix} 0 & \sin\theta \\ \sin\theta & 0 \end{bmatrix} \tag{33.2}$$

$$= \frac{\hbar}{2}\begin{bmatrix} \cos\theta & \sin\theta \\ \sin\theta & -\cos\theta \end{bmatrix}$$

This has matrix has eigenvalues $\pm\frac{\hbar}{2}$. How-so-ever we choose to orientate our axes, the eigenvalues are $\pm\frac{\hbar}{2}$. It is the coefficients, c_i, of the eigenstates within the superposition of eigenstates that change.

• Quantum Mechanics

The eigenvectors of $\widehat{S_{z\theta}}$ associated with our rotated axes are:

$$|\uparrow\rangle = \begin{bmatrix} \cos\dfrac{\theta}{2} \\ \sin\dfrac{\theta}{2} \end{bmatrix}, \qquad |\downarrow\rangle = \begin{bmatrix} -\sin\dfrac{\theta}{2} \\ \cos\dfrac{\theta}{2} \end{bmatrix} \qquad (33.3)$$

These eigenvectors are double valued. If we rotate our axes through 2π, because of the $\dfrac{\theta}{2}$ within the components of the eigenstates, the eigenstates are not identical to the ones with which we started – they are the negatives of the ones with which we started. To get back to the original eigenstates, we must rotate through 4π. This is the reason why we sometimes hear that electrons, and all other spin $\dfrac{1}{2}$ particles, must be rotated through 720° to return to where the state in which they started. This reversal of state by rotation through 2π is not conceptually understood by anyone, but it has been confirmed by experiment[1] using neutrons.

33.1 MAGNETIC MOMENTS

The "doubling" of rotational period for intrinsic spin has magnetic effects. The orbital angular momentum of an electron is associated with a magnetic moment:

$$\vec{\mu} = -\frac{e}{2m_e}\vec{L} \qquad (33.4)$$

Wherein e is the charge of the electron and m_e is the mass of the electron. This is exactly what we would expect classically. However, the electron also has a magnetic moment due to its intrinsic spin, predicted by the Dirac equation:

$$\vec{\mu_s} = -\frac{e}{m_e}\vec{S} \qquad (33.5)$$

---

[1] Greenberger. Rev. Mod. Phys, 55, 875–905 (1983).

Notice the 2 has disappeared from the denominator, and so spin has twice as much magnetic dipole as orbital angular momentum. This is inexplicable classically.

33.2   SPIN WAVEFUNCTIONS

In QFT, intrinsic spin is dealt with using Dirac's equation. However, it is possible to use Schrödinger's equation. To do this, we need to allow the wavefunction, ψ, to be a two component object:

$$\Psi(t,r) = \begin{bmatrix} \psi_1(t,r) \\ \psi_2(t,r) \end{bmatrix} \tag{33.6}$$

This is what we mean when we speak of a two-valued wavefunction. The reader might recall that we previously stated that the wavefunction has to be single valued. Well, it does, but here we are making it two-valued. Normalization is:

$$\int d\tau \left(\psi_1 \psi_1^* + \psi_2 \psi_2^* \right) = 1 \tag{33.7}$$

The Hamiltonian for spin evolving in a magnetic field is based upon the sum of the Pauli matrices:

$$\widehat{H} = -\gamma \frac{\hbar}{2} \begin{bmatrix} B_z & B_x - iB_y \\ B_x + iB_y & -B_z \end{bmatrix} \tag{33.8}$$

Wherein γ is the gyro magnetic ratio. The TDSE is:

$$i\hbar \frac{\partial \begin{bmatrix} a_1 \\ a_2 \end{bmatrix}}{\partial t} = \widehat{H} \begin{bmatrix} a_1 \\ a_2 \end{bmatrix} \tag{33.9}$$

CHAPTER 34

IDENTICAL PARTICLES

Within the macroscopic world, we can always distinguish between two similar objects. Two white billiard balls might appear to be identical at first sight, but, with enough scrutiny, we would be able to detect minute differences between them. Even if we could not detect any differences between the billiard balls, when we make them collide on a billiard table, we are able to keep track of which one is which. This is not the case with atomic particles.

Every electron in the universe is identical to all other electrons. No amount of scrutiny will detect a single difference between any two electrons. It is as if there is only one electron that is in trillions of different places at the same time. Furthermore, if we collide two electrons together, we are unable to keep track of which is which. This is not because we lack a sufficiently powerful microscope, it is because electrons have wave-like properties. Imagine a length of rope with identical wave producing machines at each end. The machines each produce a wave which travels towards the middle of the rope. When the waves collide, they become a temporary super-position of the two waves, and then the two waves separate. The waves are identical, and we ask, "Did the waves pass through each other or did they rebound from each other?". There is no way to know because the waves are identical. So it is with colliding electrons.

A state function of two electrons at $\{x_1, x_2\}$ is of the form $\psi(x_1, x_2)$. The only observable associated with this state function is the modulus $|\psi(x_1,x_2)|^2$. Because the particles are identical, the modulus of the state function for the particles being "swapped" must be equal to the modulus of the "unswapped" electrons.

$$|\psi(x_1, x_2)|^2 = |\psi(x_1, x_2)|^2 \tag{34.1}$$

This means:

$$\psi(x_1, x_2) = \psi(x_2, x_1)$$

or $\qquad(34.2)$

$$\psi(x_1, x_2) = -\psi(x_2, x_1)$$

We say that the first of these possibilities is a symmetric wavefunction, and we say that the second of these possibilities is an anti-symmetric wavefunction. It is found that these two possibilities correspond to two different types of particle. Particles with symmetric wavefunctions are called bosons, examples are the photon and the $\{W^{\pm}, Z^0\}$ bosons. Particles with anti-symmetric wavefunctions are called fermions, examples are the electron and the quarks. Fermions are said to satisfy Fermi-Dirac statistics. Bosons are said to satisfy Einstein-Bose statistics.

34.1 THE PAULI EXCLUSION PRINCIPLE

The state function for multiple different types of particles is a product of the state functions for the individual particles. For example, the state function of a photon at $x1$ and an electron at $x2$ taken together is of the form:

$$\psi_{\text{Photon}}(x_1)\psi_{\text{Electron}}(x_2) \tag{34.3}$$

To form the total wavefunction of two particles that are in the same place, $x_1 = x_2$, we need to form a linear sum of the form:

$$\Psi(x_1, x_2) = \frac{1}{\sqrt{2}}\left(\psi_{\text{P1}}(x_1)\psi_{\text{P2}}(x_2) + \psi_{\text{P1}}(x_2)\psi_{\text{P2}}(x_1)\right) \tag{34.4}$$

In which the coefficient is a normalizing constant.

If the particles are identical bosons,

$$\psi_{\text{P1}}(x_1)\psi_{\text{P2}}(x_2) = \psi_{\text{P1}}(x_2)\psi_{\text{P2}}(x_1) \tag{34.5}$$

and we have:

$$\Psi_{Bosons}(x_1,x_2) = 2\frac{1}{\sqrt{2}}\psi(x_1)\psi(x_1) \tag{34.6}$$

If the particles are identical fermions (and hence with identical wave-functions) like two identical electrons, $\psi_{P1}(x_1)\psi_{P2}(x_2) = -\psi_{P1}(x_2)\psi_{P2}(x_1)$, and we have:

$$\Psi_{Fermions}(x_1,x_2) = 0 \tag{34.7}$$

The probability of this state being observed is the modulus of this wavefunction:

$$P = |\Psi_{Fermions}(x_1,x_1)|^2 = 0 \tag{34.8}$$

We see that there is zero probability of two identical electrons, or other identical fermions, being at the same place. This is called the Pauli exclusion principle. It is essential that the electrons (fermions) are identical for the Pauli exclusion principle to apply. For example, if one electron has spin up and the other electron has spin down, the wavefunctions of these two electrons are not identical and these two electrons can be at the same place.

For bosons, there is no reason why identical bosons cannot be at the same place. For example, we can get as many photons as we like at the same place[1].

Technically, two identical electrons are two electrons which have the same set of quantum numbers. These quantum numbers include 4 numbers that are the electron's position in space-time.

The entire electron structure of atoms is based upon the Pauli exclusion principle. Because intrinsic spin, either up or down, is an electron quantum number, two electrons with different spin can occupy the same place. Electrons can, and do, form spin up and spin down pairs that share a particular orbit in an atom; "share a particular orbit" means that the electrons have the same quantum numbers regarding orbital angular momentum, energy etc. The difference of one quantum number, spin, is sufficient to make the electrons not identical. Once a pair of spin up and spin down electrons have occu-

---

[1] Presumably, this is the nature of the photon torpedos used in Star Trek.

pied an orbit, other electrons must have quantum numbers other than spin that are different from both electrons in the spin up spin down pair. These other quantum numbers are things like angular momentum, which is another way of saying that only two electrons, a spin up spin down pair, can occupy a particular atomic orbit. Electron orbitals are the basis of all chemistry as is encapsulated in the periodic table of the elements. And so you see, the periodic table of the elements is predicted by quantum mechanics – not bad eh!

34.2   THE SPIN STATISTICS THEOREM

The spin statistics theorem was first derived by Markus Fierz (1912–2006) in 1939[2] and independently proven by Wolfgang Pauli in 1940[3]. It is of central importance in quantum mechanics and quantum field theory. In a nutshell, the spin statistics theorem simply says that particles with half integral spin are fermions and particles with integral spin are bosons.

Although the Pauli exclusion principle was originally brought into physics by Pauli as a basic principle, it is not a basic principle but a consequence of the spin statistics theorem. Even so, the name "Pauli exclusion principle" is still in common usage.

---

[2] Uber die relativistische theorie kräfterfreier teichen mit beliebigem spin : Helvetica Physica Acta 12: 3–37 (1939).

[3] W. Pauli : The connection between spin and satistics : Phys Rev 58 716–722 (1940).

35

THE HYDROGEN ATOM

One of the earliest and one of the greatest achievements of quantum mechanics is the prediction of the spectrum of the hydrogen atom. The spectrum of any element is the differences between the energy levels of the electrons within the atom. Quantum mechanics, via the Pauli exclusion principle, is able not only to explain the energy levels of atoms but to also explain the whole structure of the chemical periodic table and thus the whole of chemistry. Quantum mechanics even explains the stability of DNA molecules that underpin life. To explain the hydrogen atom with perfect accuracy, in addition to quantum mechanics, we need to make minor relativistic corrections, but they are no of concern in this book. The mathematics of the hydrogen atom calculations are complicated, and we give only an outline here.

35.1 A PARTICLE IN A CENTRAL POTENTIAL

The Hamiltonian for a particle of mass, m_e, in a central potential, $V(r)$, is:

$$\widehat{H} = -\frac{\hbar^2}{2m_e}\nabla^2 + V(r) \tag{35.1}$$

The associated energy eigenvalue equation, TISE, is:

$$\left[-\frac{\hbar^2}{2m_e}\nabla^2 + V(r)\right]u_{E_n}(r,\theta,\phi) = E_n u_{E_n}(r,\theta,\phi) \tag{35.2}$$

In spherical polar co-ordinates, this is:

$$\left[-\frac{\hbar^2}{2m_e} \left(\begin{array}{c} \frac{1}{r^2}\frac{\partial}{\partial r}\left(r^2\frac{\partial}{\partial r}\right) \\ +\frac{1}{r^2\sin\theta}\frac{\partial}{\partial \theta}\left(\sin\theta\frac{\partial}{\partial \theta}\right) \\ +\frac{1}{r^2\sin^2\theta}\frac{\partial^2}{\partial \phi^2} \end{array} \right) + V(r) \right] u_{E_n}(r,\theta,\phi) \qquad (35.3)$$

$$= E_n u_{E_n}(r,\theta,\phi)$$

Looking at the total orbital angular momentum operator, $\widehat{L^2}$, we see that (35.3) can be written as:

$$\left[\begin{array}{c} -\frac{\hbar^2}{2m_e}\frac{1}{r^2}\frac{\partial}{\partial r}\left(r^2\frac{\partial}{\partial r}\right) \\ +\frac{1}{2m_e r^2}\widehat{L^2} + V(r) \end{array} \right] u_{E_n}(r,\theta,\phi) = E_n u_{E_n}(r,\theta,\phi) \qquad (35.4)$$

Looking at this, we see that, for a central potential, the Hamiltonian commutes with both $\widehat{L^2}$ and L_z. Because these three operators commute, it is possible to know the energy, total orbital angular momentum, and the z-component of the orbital angular momentum simultaneously. We denote the eigenvectors of the two angular momentum operators with the letters $\{l, m\}$. Since the $\{\phi, \theta\}$ dependence of the Hamiltonian is entirely within the $\widehat{L^2}$ term, we have:

$$u_{nlm}(r,\theta,\phi) \quad u_{nl}(r)Y_l \ (\theta,\phi) \qquad (35.5)$$

Wherein the $Y_l^m(\theta,\phi)$ are the spherical harmonics we dealt with above. We are thus able to write the Hamiltonian for a central potential as:

$$\left[\begin{array}{c} -\frac{\hbar^2}{2m_e}\frac{1}{r^2}\frac{\partial}{\partial r}\left(r^2\frac{\partial}{\partial r}\right) \\ +\frac{l(l+1)\hbar^2}{2m_e r^2} + V(r) \end{array} \right] u_{nl}(r) = E_n u_{nl}(r) \qquad (35.6)$$

Aside: With enough algebraic manipulation, the Hamiltonian with a central potential, $V(r)$, can be cast in the form:

$$\widehat{H} = \frac{p_r^2}{2m_e} + \frac{p_t^2}{2m_e} + V(r) \tag{35.7}$$

Wherein p_r is the momentum in a radial direction and p_t is the momentum in the transverse direction.

35.2 THE HYDROGEN ATOM

For an atomic nucleus, the Coulomb potential is:

$$V(r) = -\frac{Ze^2}{r} \tag{35.8}$$

Where Z is the number of protons in the nucleus.

We put:

$$E_n = -|E_n|, \qquad \alpha_n^2 = \frac{8m_e|E|}{\hbar^2}$$

$$\rho = \alpha_n r, \qquad \lambda_n = \frac{Ze^2}{\hbar}\left(\frac{m_e}{2|E_n|}\right)^{\frac{1}{2}} \tag{35.9}$$

And we get:

$$\left[\frac{1}{\rho^2}\frac{\partial}{\partial\rho}\left(\rho^2\frac{\partial}{\partial\rho}\right) + \frac{\lambda_n}{\rho} - \frac{1}{4} - \frac{l(l+1)}{\rho^2}\right]u_{nl}(\rho) = 0 \tag{35.10}$$

There are singularities at $\{\rho = 0, \rho = \infty\}$, and we are seeking the eigenvalues λ_n. A solution is:

$$u_{nl}(\rho) = \rho^s e^{-\frac{\rho}{2}} L_{nl}(\rho) \tag{35.11}$$

Wherein

$$L_{nl}(\rho) = \sum_\kappa a_\kappa \rho^\kappa \tag{35.12}$$

This series must terminate to provide a solution. This leads to the energy levels of the electrons being given by:

$$E_n = -\frac{1}{2}\frac{Z^2 e^2}{a_0}\left(\frac{1}{n^2}\right) \tag{35.13}$$

Wherein $a_0 = \dfrac{\hbar^2}{m_e e^2}$ is the Bohr radius. For hydrogen, $Z = 1$, and we have the electron energies of hydrogen are:

$$E_{Hydrogen} = -\frac{1 m_e e^4}{2\hbar^2 n^2} \quad : \quad n = 1,2,3,... \tag{35.14}$$

35.3 ELECTRON ORBITS

The energy levels are determined by a single quantum number, n. This is called the principle quantum number. For each energy level, there are n values of orbital angular momentum corresponding to the orbital quantum numbers $l = 0, 1, 2, ...(n-1)$. $l = 0$ corresponds to a straight line oscillation through the nucleus. Corresponding to each value of l, there is a total angular momentum eigenvalue of $\hbar^2 l(l+1)$. For each value of l, there 2l + 1 values of the magnetic quantum number, m, ranging as $m = 0, \pm1, \pm2,...\pm l$ Corresponding to each value of m, there is a l_z eigenvalue of $m\hbar$. Although the calculation above does not lead to it, there is associated with each orbit also a spin quantum number which is either up or down.

Each set of the quantum numbers $|n,l,m\rangle$ determines a particular eigenfunction. There are thus a number of different eigenfunctions for each energy level. This is called degeneracy. Not counting spin, there are n^2 eigenstates for each value of n; counting spin, there are $2n^2$ eigenstates for each value of n.

The four quantum numbers $|n,l,m,s\rangle$ and the Pauli exclusion principle determine the chemistry and structure of the chemical elements as portrayed in the periodic table – that is not a small achievement for a theory. There is a lot of difference between the way we have deduced the structure of the atoms above and the way that Bohr arrived at almost the same place. Bohr arrived here by adding guessed *ad hoc* rules to classical mechanics. We arrived here by building a mathematical structure completely independently of classical mechanics. Our mathematical structure, in spite of it being profoundly different from the mathematical structure of classical mechanics, subcludes within it classical mechanics in macroscopic systems. This mathematical structure is called quantum mechanics.

EXERCISES

1. What type of functions are used to write the angular part of the wavefunction of the hydrogen atom?

2. What is the energy of the lowest ($n = 1$) electron orbit in the hydrogen atom – give your answer in electron volts?

CHAPTER 36

RELATIVISTIC QUANTUM MECHANICS

We moved from Newtonian mechanics to quantum mechanics by forming a quantum mechanical operator corresponding to every Newtonian dynamic variable except time such that the classical variables become quantum operators. If we take Maxwell's theory of electromagnetism and do the equivalent change of dynamic variables into operators, then, because Maxwell's theory is a relativistic theory, we will get a relativistic form of quantum mechanics. The quantum theory of radiation derived this way precisely confirms Planck's hypothesis that radiation may be considered as a collection of particles of zero mass (photons) with energy and frequencies determined as he hypothesized.

When these photons interact with electrons there is a probability that the electron will absorb or emit a photon. The size of that probability of photon emission or photon absorption by electrons determines the strength of the electromagnetic force as expressed in the fine structure constant:

$$\alpha = \frac{e^2}{\hbar c} = \frac{1}{137} \tag{36.1}$$

Having constructed a relativistic quantum theory of radiation, we need to modify the Schrödinger equation for the hydrogen atom so that it too is consistent with the special theory of relativity. This was done by Dirac in 1928[1]. The consequences of imposing special relativistic requirements upon quantum theory leads to:

a. Electrons have intrinsic spin of $\frac{\hbar}{2}$.

b. Spin will have double the magnetic moment of orbital momentum as expressed in

$$\vec{\mu}_s = -\frac{e}{m_e}\vec{S} \qquad (36.2)$$

c. There is an anti-particle to the electron - the positron.

d. The energy levels of the hydrogen atom are changed slightly in a way that exactly agrees with experimental observations.

e. There will be a "back reaction" on the electron from radiation due to its orbital acceleration. This "back reaction" results in fine splitting of the electron energy levels. This fine splitting was measured by Willis Lamb (1913–2008) in 1947[2]; it is known as Lamb shift and the predictions exactly fit the observations. The calculations associated with the magnetic moment and other properties of the electron and the Lamb splitting are very complicated and are done by computer.

To date, using relativistic quantum mechanics, the magnetic moment of the electron has been calculated and verified by experiment to twenty-eight decimal places.

---

[1] P.A.M.Dirac Proc. Roy. Soc. A117, 610 (1928).
[2] Lamb, Willis, E, Retherford, Robert, C. (1947) "Fine structure of the hydrogen atom by a microwave method" Physical Review 72(3) 241–243.

CONCLUDING REMARKS

This book is no more than an introduction to quantum mechanics. I hope this book has provided the reader with a solid basis upon which to build a deeper knowledge of this very successful area of human endeavor.

I ended the book with the successful prediction of the periodic table of the elements, the successful prediction of the spectra of hydrogen, and the extremely successful prediction of the magnetic moment of the electron. With such successes, it would seem that the theory of quantum mechanics is beyond question. It is not. Humankind does not properly understand the "weird" bits of quantum mechanics such as superposition, collapse of the wavefunction, non-locality and entanglement. There are many often unasked questions about the universe which quantum mechanics does not answer:

1. Why three Pauli matrices? Why not four or two? The three Pauli matrices can be written as six symmetric 4×4 matrices. There are only six possible such matrices that are the square roots of plus one.

2. What is spin? Angular momentum means different things in different types of space. Perhaps the existence of spin means we need to revise our concepts of space and time.

3. Why wave-particle duality? Why not a triality? We seem to live in a space with six 2-dimensional rotation planes. These six rotations are of two types, space-time and space-space. Perhaps this is why we have a duality of existence rather than the monality we might expect.

4. What is mass? Although the Higgs mechanism proffers an explanation of the origin of mass, the Higgs mechanism is

seen by many theoretical physicists as messy and not beautiful enough to be true.

5. What is electric charge? We speak of electric charge as if we know of what we speak, but what electric charge really is we have not even a clue.

6. Where does gravity fit into quantum mechanics? There's a Nobel prize for answering this question.

7. What is space and time and how does empty space expand in the universe?

8. What happens at the very high energies of the big bang?

9. Why the particular particle content of the universe?

10. How many physical constants are there, and why are they the values they are? We have seen that physical constants might be scaling parameters in different division algebras, but this research is still in its infancy.

11. Why does nature have four forces instead of three or just one? Perhaps there are more forces at higher energies, super-symmetric forces perhaps.

12. Is the Dirac function really a sensible thing?

Humanity's quest to understand the universe is far from complete. Perhaps the reader will go on to deepen this understanding.

I hope the reader has been sitting comfortably through the whole of this book. I recommend that the reader now re-reads the book twice more; you will sit even more comfortably through the second reading than you did through the first.

APPENDIX

THE POSTULATES OF QUANTUM MECHANICS

This is a set of six postulates of quantum mechanics. From these, we can deduce all the consequences that, together with these postulates, are called the theory of quantum mechanics. It is possible to swap postulates for consequences, and other authors might give a slightly different set of postulates, but this set is the most commonly accepted set.

POSTULATE 1

The state of a physical system at time t is represented by a complex wavefunction, $\Psi(t, x)$, such that the state is normalized:

$$\int_{-\infty}^{\infty} dx \; \Psi^*\Psi = 1 \tag{A.1}$$

We sometimes see this presented as: a particle moving in a potential (might be zero) is associated with a wavefunction. The two states $\Psi(t,x)$ & $e^{i\alpha}\Psi(t,x)$ are seen by the observer to be the same state.

The unobserved state, $\Psi(t, x)$ is a superposition (linear sum with complex coefficients) of all the possible observed states, ψ_i.

$$\Psi = c_1\psi_1 + c_2\psi_2 + c_3\psi_3 + ... \tag{38.2}$$

The observed state will be one of the possible observed states, ψ_i. The wavefunction, $\Psi(t, x, y, z)$, is a single valued complex function of the space and time co-ordinates. The modulus squared of this wavefunction, $\Psi^*\Psi$, evaluated at a particular point in space and time is the probability density of the wavefunction at that particular point in space and time.

POSTULATE 2

A linear Hermitian operator corresponds to each observable property (energy, momentum,...) of a physical system. The possible outcomes of measuring the observable property are the eigenvalues of the operator. The commutators of these operators are either zero or proportional to $\hbar$.

POSTULATE 3

The momentum operator is $\widehat{p_x} = -i\hbar \dfrac{\partial}{\partial x}$. The position operator is $\hat{x} = x$. Most other operators follow from this by copying the relations between Newtonian dynamic variables. An exception is intrinsic spin.

POSTULATE 4

The wavefunction, Ψ, is a linear sum of eigenfunctions, ψ_i, that are solutions to the time independent Schrödinger equation. Ψ is normalized. If $\Psi = c_1\psi_1 + c_2\psi_2 + c_3\psi_3 + \ldots$, then the probability of the state ψ_i being observed is given by $P = |c_i|^2$..

POSTULATE 5

The result of a measurement can be predicted with certainty only if the physical system is in a particular eigenstate, ψ_i, of the system, otherwise the result is a matter of chance.

POSTULATE 6

Associated with each physical system is a linear Hermitian energy operator, $\widehat{H}$, which determines the time evolution of the physical system through Schrödinger's time dependent equation $i\hbar\dfrac{\partial\Psi}{\partial t} = \widehat{H}\Psi$. We often see this described as "The wavefunction satisfies the Time dependent Schrödinger equation at all times."

A List of Physical Constants
Bohr magneton

$$\mu_B = \frac{e\hbar}{2m_e}$$ 9.274×10^{-24} $\mathrm{J\,T^{-1}}$

Bohr radius

$$a_0 = \frac{4\pi\varepsilon_0\hbar^2}{m_e e^2}$$ 5.292×10^{-11} M

Boltzmann's constant

$$k_B$$ 1.3807×10^{-23} $\mathrm{J\,K^{-1}}$

Electron charge

$$e$$ 1.602×10^{-19} C

Electron mass

$$m_e$$ 9.109×10^{-31} Kg

Electron wavelength

$$\lambda_0 = \frac{h}{m_e c}$$ 2.426×10^{-12} M

Planck's constant

$$h \qquad\qquad \begin{cases} 6.626\times10^{-34} & \text{J S} \\ 4.135\times10^{-15} & \text{eV} \end{cases}$$

Aitch bar

$$\hbar = \frac{h}{2\pi} \qquad\qquad 1.0546\times10^{-34} \quad \text{J S}$$

Speed of light in vacuo

$$c \qquad\qquad 2.998\times10^{8} \quad \text{M S}^{-1}$$

Conversion Factors

$$1 \text{ eV} = 1.602\times10^{-19} \quad \text{J}$$
$$1 \text{ Mev} = 1.78\times10^{-30} \quad \text{Kg}$$
$$1 \text{ Å} = 10^{-10} \text{ M}$$

Useful Formulae

$$\int_{-\infty}^{\infty} dx\, e^{-ax^2} = \sqrt{\frac{\pi}{a}}$$

BIBLIOGRAPHY

Davis, Paul C. W. & Betts, David S. Quantum Mechanics, ISBN: 0-412-57900-6

Feynman, R.P. & Leighton, R.B. & Sands, M., The Feynman Lectures on Physics Volume III, ISBN: 0-201-02118-8

French, A.P. & Taylor, E.F., An Introduction to Quantum Physics, ISBN: 0-412-37580-X

Greenhow, R.C. Introductory Quantum Mechanics, ISBN: 0-7503-0010-8

McMahon, David, Quantum Mechanics Demystified, ISBN: 0-07-145546-9

Sudbery, Anthony, Quantum Mechanics and the Particles of Nature, ISBN: 0-521-27765-5

Weinberg, Steven, Lectures on Quantum Mechanics, ISBN: 978-1-107-02872-2

Zee, A, Quantum Field Theory, ISBN: 0-691-01019-6

BOOKS BY DENNIS MORRIS

Morris, Dennis, Complex Numbers, The Higher Dimensional Forms, ISBN: 978-1508877499

Morris, Dennis, Empty Space is Amazing Stuff, ISBN: 978-0-9549780-7-5

Morris Dennis, The Physics of Empty Space, ISBN: 978-1507707005

Morris Dennis, The Naked Spinor, ISBN: 978-1507817995

INDEX